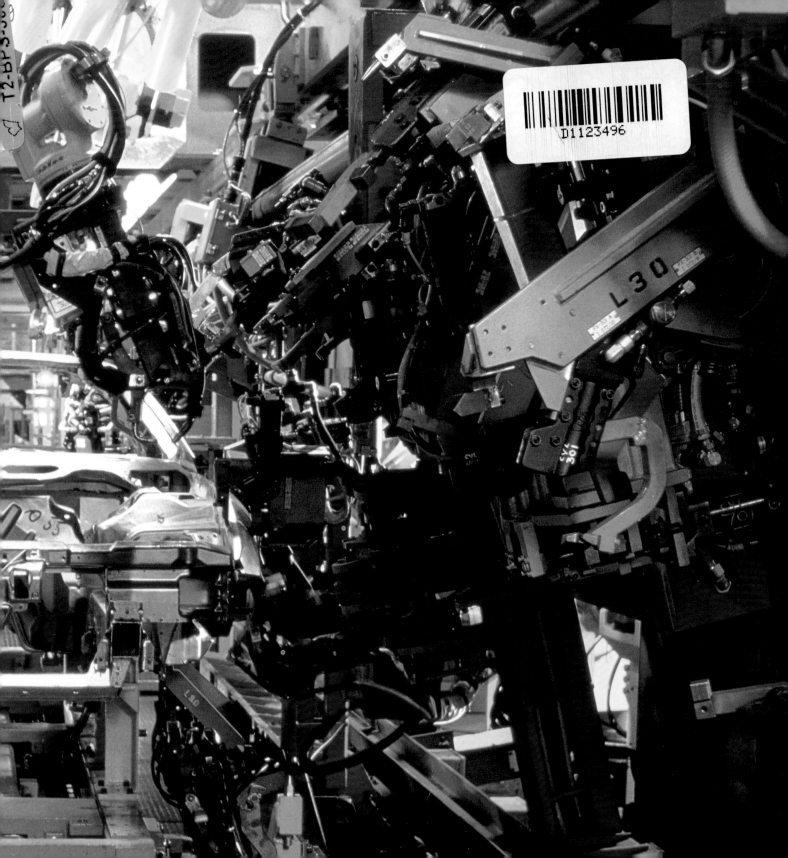

◄ Inside a nuclear fusion reactor where research is in progress. The bright lines are electrical discharges that have been generated by an immensely powerful machine called a particle beam fusion accelerator.

◁ Overleaf An automated car production line consists entirely of machines run by machines. Assembly starts with a bare chassis (frame), and parts are added on as the car moves along the conveyor.

inventions

explained

A Beginner's Guide to Technological Breakthroughs

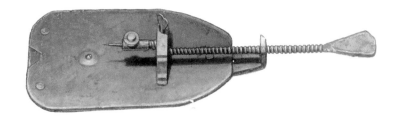

RICHARD PLATT

Henry Holt and Company
New York

Henry Holt and Company, Inc.
Publishers since 1866
115 West 18th Street
New York, New York 10011

Henry Holt® is a registered trademark
of Henry Holt and Company, Inc.

Published in Canada by Fitzhenry & Whiteside Ltd.,
195 Allstate Parkway, Markham, Ontario L3R 4T8.

Library of Congress Cataloging-in-Publication Data
Platt, Richard.
Inventions explained: a beginner's guide to technological
breakthroughs Richard Platt.
p. cm.— (A Henry Holt reference book) (Your world
explained series)
Includes index.
ISBN 0-8050-4876-6 (alk. paper)
1. Inventions — History. I. Title. II. Series III. Series:
Your world explained series.
T15.P625 1997
609 — dc21   97-11969
                                                  CIP

Henry Holt books are available for special promotions
and premiums. For details contact: Director, Special Markets.

First Edition—1997

Editor:	Catherine Baxter
Designers:	Steve Woosnam-Savage, Siân Williams
Art Director:	Ralph Pitchford
Managing Editor:	Kate Phelps
Editorial Director:	Cynthia O'Brien
Production:	Janice Storr, Selby Sinton
Research:	Lynda Wargen
Picture Research:	Zilda Tandy, Sue Alexander

Printed and bound in Portugal by Printer Portuguesa
Originated in Singapore by Master Image
All first editions are printed on acid-free paper.

10 9 8 7 6 5 4 3 2 1

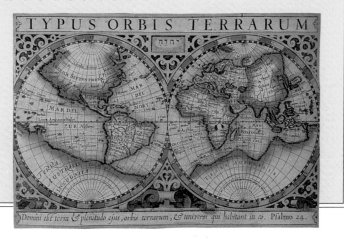

Contents

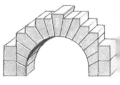

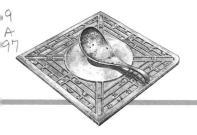

INTRODUCTION

THIS BOOK TELLS THE FASCINATING STORY OF INVENTION. The tale begins with the invention of tools in Africa and more than two million years of human ingenuity unfold in the pages that follow.

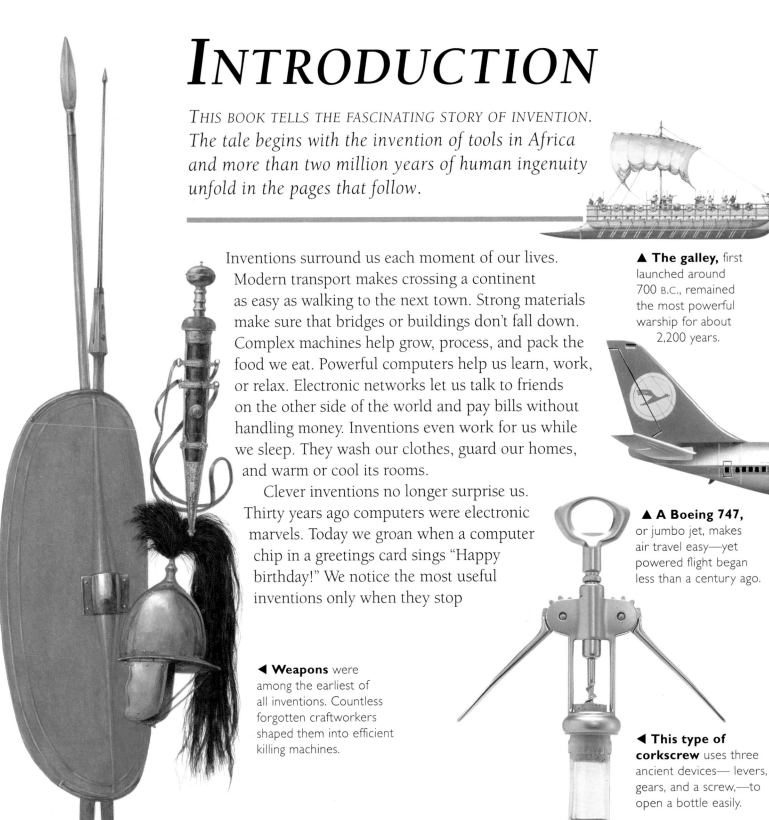

Inventions surround us each moment of our lives. Modern transport makes crossing a continent as easy as walking to the next town. Strong materials make sure that bridges or buildings don't fall down. Complex machines help grow, process, and pack the food we eat. Powerful computers help us learn, work, or relax. Electronic networks let us talk to friends on the other side of the world and pay bills without handling money. Inventions even work for us while we sleep. They wash our clothes, guard our homes, and warm or cool its rooms.

Clever inventions no longer surprise us. Thirty years ago computers were electronic marvels. Today we groan when a computer chip in a greetings card sings "Happy birthday!" We notice the most useful inventions only when they stop

▲ **The galley,** first launched around 700 B.C., remained the most powerful warship for about 2,200 years.

▲ **A Boeing 747,** or jumbo jet, makes air travel easy—yet powered flight began less than a century ago.

◄ **Weapons** were among the earliest of all inventions. Countless forgotten craftworkers shaped them into efficient killing machines.

◄ **This type of corkscrew** uses three ancient devices— levers, gears, and a screw,—to open a bottle easily.

8

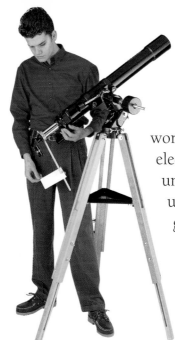

▲ The modern telescope is almost unchanged since its invention in 1608.

▲ Tanks, which were invented in 1915, changed the way that armies fought wars.

working, and we take electricity for granted until a black out reminds us just how much we have grown to depend on it.

A few of these inventions are the brilliant ideas of famous scientists, but many more are the result of centuries of slow improvement by a number of different people. Nobody now remembers the craftworkers who perfected ancient tools such as the ax, or weapons such as bows, spears, and swords.

Today scientists and engineers work in large groups. Together they create smart new machines which make our lives better. Every invention makes the next one a little easier as each generation of inventors learns about what has gone before, and gains the wisdom needed to create new inventions for the future.

The inventors of the past have achieved a lot. But the world still urgently needs bright new ideas. We need modern cures for disease. We need to clean up the environment. We need to feed and house more people each year. Will you be the inventor who solves these growing problems?

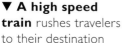

▼ A high speed train rushes travelers to their destination with far less pollution than cars.

▲ The compass, a Chinese invention, was first used as a novelty north-pointing spoon.

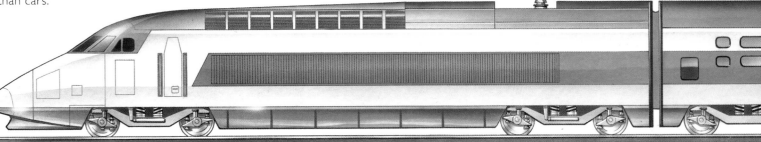

ANCIENT INVENTORS SHAPE THE WORLD

IN THEIR STRUGGLE TO SURVIVE, THE WORLD'S FIRST HUMANS created new tools and methods of hunting, clearing land, building houses, and growing food. Later inventions made travel faster, and started the first industries.

Tools and weapons

The earliest human beings lived in eastern Africa about 2.5 million years ago. They had large brains and were more inventive than their ancestors, the apes. They made tools and weapons from stone, wood, and bone. Stone provided the only truly sharp cutting edges until 5,500 years ago, when people in the Middle East discovered how to make tools with metal. Smiths hammered long sharp swords from copper and bronze. Warriors fighting with these new superior swords easily beat enemies armed only with stone weapons.

▲ **An otter deftly** cracks a shell with an anvil stone, and shows that humans are not the only animals to use tools.

Before metal
The Stone Age is so called because stone tools have survived long after other traces of people have decayed. The first stone tools were crude, but they were sharp enough to shave wooden poles into deadly spears for hunting. Over thousands of years, people found better ways to flake stone. They made it into all sorts of different tools, and eventually they learned to sharpen and polish the edges.

These skills are used to divide the Stone Age into two periods. *Paleolithic* (old Stone Age) people made chipped-edge tools. *Neolithic* (new Stone Age) people made polished tools.

crude tool

10

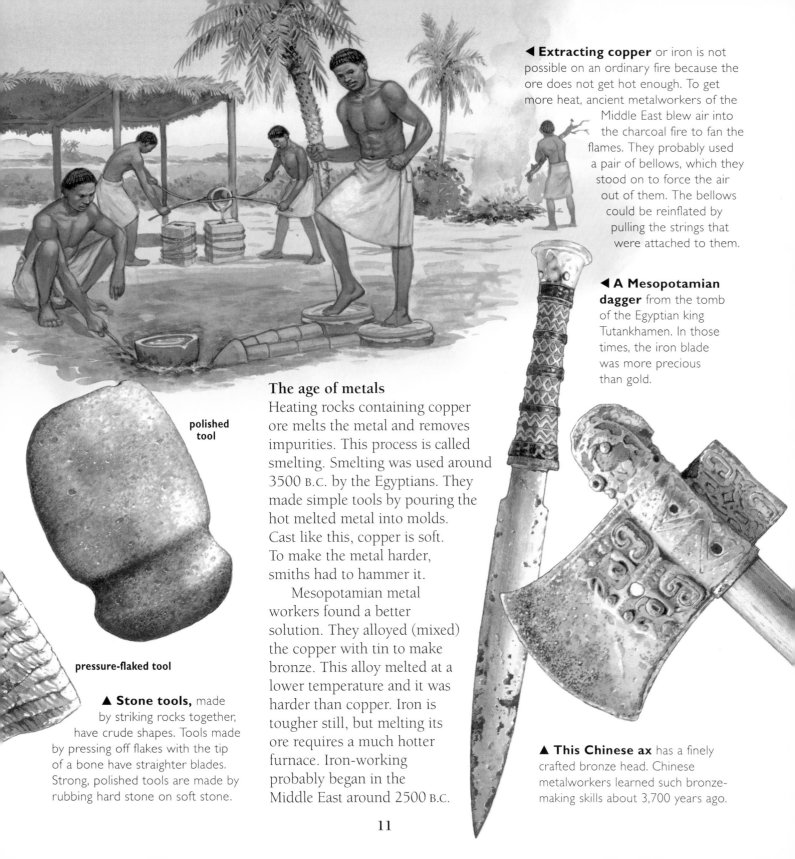

◄ Extracting copper or iron is not possible on an ordinary fire because the ore does not get hot enough. To get more heat, ancient metalworkers of the Middle East blew air into the charcoal fire to fan the flames. They probably used a pair of bellows, which they stood on to force the air out of them. The bellows could be reinflated by pulling the strings that were attached to them.

◄ A Mesopotamian dagger from the tomb of the Egyptian king Tutankhamen. In those times, the iron blade was more precious than gold.

polished tool

pressure-flaked tool

▲ Stone tools, made by striking rocks together, have crude shapes. Tools made by pressing off flakes with the tip of a bone have straighter blades. Strong, polished tools are made by rubbing hard stone on soft stone.

The age of metals

Heating rocks containing copper ore melts the metal and removes impurities. This process is called smelting. Smelting was used around 3500 B.C. by the Egyptians. They made simple tools by pouring the hot melted metal into molds. Cast like this, copper is soft. To make the metal harder, smiths had to hammer it.

Mesopotamian metal workers found a better solution. They alloyed (mixed) the copper with tin to make bronze. This alloy melted at a lower temperature and it was harder than copper. Iron is tougher still, but melting its ore requires a much hotter furnace. Iron-working probably began in the Middle East around 2500 B.C.

▲ This Chinese ax has a finely crafted bronze head. Chinese metalworkers learned such bronze-making skills about 3,700 years ago.

11

Fire, power, and simple machines

▲ **When building a canoe,** Central American people of the 16th century used flames to hollow a log. Chinese people were the first to make use of fire for cooking and keeping warm.

Primitive humans discovered fire about half a million years ago, probably when lightning set a tree ablaze. However, it was a long time before people learned that fire could be used to burn down trees deliberately to clear land for farming. Later, people looked for other ways to make their lives easier. They tamed cattle to pull loads and, about 2,000 years ago, people began to use water power. The use of simple machines made people even more efficient.

water chute

grain

hopper

turning millstone

stationary millstone

▶ **Water-driven grain mills** used the force of a rushing stream to spin a simple turbine. This turned the millstones above, which ground coarse grain into fine flour.

trough for collecting flour

turbine

Power to change society

Fire cleared fields for planting crops, which meant that nomadic (wandering) people could become farmers and settle in larger groups. Many of these new societies fought their neighbors for water and land. They took prisoners to work as slaves, thus making their own lives increasingly comfortable.

Oxen began to take the place of slaves in providing power 5,000 years ago in the Middle East. Buffalo, elephants, and camels soon pulled and carried in other regions.

Pulleys, Winches, and Cranes

Hauling a heavy bucket of water up a deep well is exhausting. It's not so hard though, if the rope lifting the bucket is passed over a pulley. The pulley changes the rope's direction, so pulling it is much easier. Passing the rope through a second pulley halves the effort needed to pull the bucket up the well, and adding another pulley (right) cuts the effort needed to a third.

Greek engineers invented pulleys 2,500 years ago and made the first winches and capstans (rope-pulling drums). Mariners used these on their ships. In Greek theaters the same devices lowered "gods" on to the stage (left).

pulley

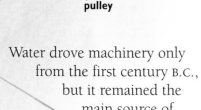

winch

Water drove machinery only from the first century B.C., but it remained the main source of power until the invention of the windmill 1,000 years later.

The most common use of water power was in turbine mills (see far left) where it was used for grinding grain to make flour. A later mill wheel, invented in the first century, could do the work of 40 slaves.

Using machines

Like pulleys (see above), slopes and wedges make work easier by spreading it out over a longer distance. The slope is the simplest. Rolling a boulder up a slope gradually lifts it to the top. Though the boulder is heavy, one person can move it. Lifting the boulder straight upward to the same height would need the strength of about 10 men. A wedge splits a log the same way, turning a small downward force into a much bigger sideways force.

▲ **The shaduf** is a water-lifting lever invented around 1500 B.C. It has a weight on one end and a bucket on the other.

Levers also can be used to make work lighter and easier. They act in the same way as seesaws: pushing one end down lifts the other. If the lever is built to pivot around its center, lifting a weight on one end requires equal downward force on the other. But if the weight is closer to the pivot, then less force is needed.

▼ **The plow uses two wedges** to prepare fields for seeding. One, at its tip, makes a gulley in the soil. The other, behind, turns the soil over, which buries weeds.

Making and measuring

As farmers found better ways to grow crops and herd animals, fewer people worked in the fields. Some became priests, warriors, or officials; others started the first industries. These craftworkers made baskets and fabrics from fibers. They made containers from pottery and glass, and beautiful ornaments from precious metals. When merchants exchanged these craft goods for raw materials, or crops and animals, they needed to estimate the weight or length of them, so weighing and measuring began. Other trades used measurement, too. Builders, for instance, had to judge distances accurately.

Roman glassware

▲ **These coiled pots from South America** are made using the same techniques that ancient potters used.

▼ **A duck-shaped weight** has its size marked on the side. Traders from Mesopotamia used this weight 2,800 years ago.

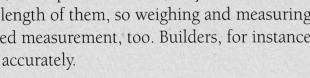

Crafting pots and glass
Damp clay is malleable and so it is easy to mold it into sculptures or bowls. Drying hardens them until the clay gets wet again. By about 7000 B.C., the people of Turkey and Iran had learned that baked clay vessels do not soften in water. This discovery enabled them to make longer-lasting pots, cups, and jars.

All were porous (they soaked up water). However, potters found they could make their wares waterproof with glaze, a glass-like coating. Making containers entirely of glass was more difficult because it required a much hotter fire. People from Mesopotamia and Egypt made the first glass vessels 3,500 years ago.

▼ **The Nile River** floods each year, hiding fields such as these under mud. Land measurement began as people had to renew the field borders regularly.

Weighing and measuring

Measuring answers the question, "How much?" by comparing everything with a standard unit— a weight or length that's the same everywhere. Length measurements are very ancient. Originally, people used their forearms as rulers. Egyptian people called this length the cubit. The distance from the elbow to the forefinger-tip varies from person to person, making cubits unreliable. Around 3000 B.C. the country began to use a standard cubit of 20.6 inches. Goldsmiths from the Indus River valley (now in Pakistan) made the first weights. They used them to weigh gold dust 4,500 years ago.

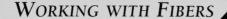

Weaving, braiding, and coiling turn threads of animal hair and plant fiber into warm blankets, useful ropes, fishing nets, or baskets. Fabrics and baskets found in Iraq and Egypt date from 5000 B.C., but these crafts probably began earlier. We may never know their true origins because, unlike pottery, fibers are quick to rot. Fabric-making is a two-stage process. Spinning twists fibers together to make a thread. Weaving on a loom turns them into cloth.

an Aztec woman weaving on a ground loom

a piece of patterned fabric like this would have taken a long time to weave on a ground loom

▼ **Winnowing grain** separates the lighter chaff (the shell) from the edible seeds. With a basket, farmers could throw the grain in the air and let the wind blow the chaff away.

▲ **Egyptian rulers** divided the cubit into shorter measures, also based on the body.
One cubit equalled seven palms. Each palm was four "fingers" wide.

merkhet

◄▼ **The merkhet,** made from a palm leaf, allowed Egyptian surveyors to set out straight lines by gazing through its splits. The cross-shaped groma ensured corners were square, and plumb lines kept surfaces level.

groma

plumb line

15

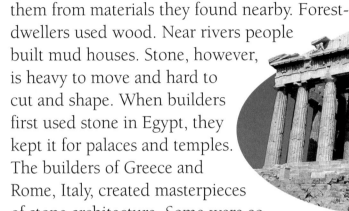

Building

Nomadic people of the past lived in simple shelters. Their homes were quick to build, or easy to pack up and carry. But as people settled and farmed, they wanted more solid houses. They built them from materials they found nearby. Forest-dwellers used wood. Near rivers people built mud houses. Stone, however, is heavy to move and hard to cut and shape. When builders first used stone in Egypt, they kept it for palaces and temples. The builders of Greece and Rome, Italy, created masterpieces of stone architecture. Some were so solid that they are still standing today.

▲ **Teepees** became the portable homes of the North American plain tribes about 2–300 years ago. Their ancestors had been using other tents for over 1,500 years.

▼ **The stilt houses** of Malaysia have developed over thousands of years. In a hot climate it is important to keep cool and the stilts help to catch the breeze. The wooden walls and thatched roof also keep the heat at bay.

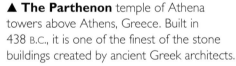

▲ **The Parthenon** temple of Athena towers above Athens, Greece. Built in 438 B.C., it is one of the finest of the stone buildings created by ancient Greek architects.

Development of building materials

When people first settled in towns in the Middle East, they probably lived in reed huts plastered with mud. Building in mud bricks may have begun on the banks of the Jordan River some 8,000 years ago. These mud bricks cracked in the sun, so Egyptian builders made them stronger by mixing straw or animal hair with the mud. Mesopotamian people made another improvement. They burned the bricks in a fire. This made them stronger and waterproof. Roman engineers also used "fired" bricks.

Concrete

The Romans also spread knowledge of concrete. However, one of its ingredients, cement, was in use much earlier. Since 2500 B.C. builders have been burning limestone to make this powder. Cement hardens when it is wet and will glue together stone and brick walls. Roman builders added ash from Mount Vesuvius, a volcano that is close to Naples in Italy.

When poured into molds, the new concrete set as hard as stone. With it, Roman engineers constructed the huge, wide arched roof for the Pantheon in Rome. It is 142 feet wide!

Climate and buildings

People in every region of the world have invented their own special building styles to suit local conditions. In the southwestern U.S.A., for example, the days are hot and the nights are cold. Here, traditional adobe houses have thick mud-brick walls and small windows. They stay cool in the day by absorbing the sun's heat, and this stored heat keeps them warm through the night. By contrast, in southeast Asia, where it is stuffy and humid, people make the most of any cooling breezes by using forest trees to build raised houses with huge windows.

▼**Making sun-dried bricks** is quick and easy. These men are using methods that have hardly changed in 6,000 years.

THE ARCH

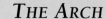

primitive arch

Egyptian arch

Roman arch

Doorways and windows weaken walls, so ancient builders supported the wall above them with a lintel—a flat beam. But lintels are also weak, so doorways had to be narrow. The invention of the arch made wider openings possible. Egyptians and Mesopotamians made small arches from 3000 B.C. Romans built huge arches by supporting them on a wooden frame.

Transport on land and sea

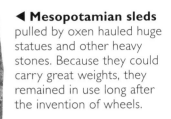

◄ **Mesopotamian sleds** pulled by oxen hauled huge statues and other heavy stones. Because they could carry great weights, they remained in use long after the invention of wheels.

Until 5,500 years ago, only footprints marked the routes of the world's paths. Then suddenly a new kind of track appeared: a pair of lines cut deep in the sand or mud. Wheels left these strange new marks. Wheeled vehicles enabled people to move heavy loads farther and travel faster. Transport on water was quicker still and ships carried even heavier cargoes. Blown along by the wind, sailing ships enabled people to trade or settle in distant lands.

The sled and the wheel

Heavy loads are exhausting to carry. So from about 4000 B.C. people used animals to do this tiring work. The donkey was probably the first pack animal. If the ground is slippery, dragging a load is easier than carrying it. Sleds are vehicles made for dragging. They work by supporting the weight on narrow runners that slide easily. People first used sleds 7,000 years

ago on northern Europe's cold, icy ground, but sleds don't need to have snow or ice to be useful. They slide extremely well on grass or mud, too.

We don't know who first put wheels on a sled, but clay tablets created around 3500 B.C. in the Sumerian town of Erech (now in Iraq) show the

◄ **The first wheels** were made from planks that were strengthened at the center. Slices from tree trunks made poor wheels because they split.

▶ **War chariots** allowed the Ancient Egyptian armies to attack their enemies at a terrifying speed. Each carried a driver and a warrior armed with a bow and arrow.

▼ An Inuit sled is made to a design that has changed little in 700 years. They were probably pulled by dogs then as they are today.

◄ Boats on Lake Titicaca, Peru, are still made from tightly bound bundles of reeds. Prehistoric Nile boats in ancient Egypt probably looked very similar to these.

▼ Lapeta pottery has been found on many Pacific islands—evidence that people made long ocean voyages 3,600 years ago.

Water transport

Boats and ships were not the first water transport. Floats of wood, reeds, or air-filled animal skins were earlier inventions. Tying several together created a long raft. It carried more weight and was easier to control than a float.

Hollowing out a raft made the first boat. Pacific island people probably made boats 40,000 years ago. Nothing remains of these boats, but we know that they existed because they spread their makers' way of life over the Pacific region.

The first real records of sailing ships come from Egypt and Mesopotamia. Fishermen, or ferrymen, on Egypt's Nile River tied twin masts to reed boats some 5,500 years ago. By raising a square sail they caught the wind, which blew them south. To travel the other way, they simply lowered the sail and drifted with the current.

first cart. The cart's wheels were heavy and probably made from planks of wood, which were joined together and sawed into circles.

The first light wheels, made around 2000 B.C., looked like wooden hoops. Spokes—thin, light rods—supported the hooplike rim. The spokes met at the center. Here, a hub held the wheel to the vehicle. These spoked wheels, when combined with light chariots and onagers (strong, primitive horses), made the first fast vehicles.

▲ Pacific island fishing boats had outriggers (balancing floats) to help keep them upright in the water. Larger oceangoing sailing canoes had twin hulls, two masts, and were up to 80 feet long.

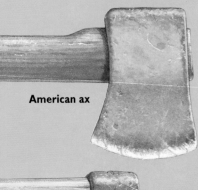

American ax

European ax

▲ **The 18th-century American ax** shows how blacksmiths continued improving tools long after the Middle Ages. The new design was easier to use than older European axes.

ARTISANS AND INVENTORS

INVENTIONS WERE NOT ALL DRAMATIC DISCOVERIES BY FAMOUS SCIENTISTS. Sometimes they began as the ideas of artisans (craftworkers) whose names have long since been forgotten. Constant improvement over thousands of years has made these inventions work perfectly.

The guild system

Today, the word "masterpiece" does not mean quite the same thing as it did in the late 15th century. Then, a masterpiece was an especially fine piece of craftwork that a journeyman (trainee worker) created to show off his skills. Those who passed this test became masters (experts) in a guild. Guilds were organizations in charge of each industry or trade. They taught the "secrets" (skills) of the crafts. But guilds often stifled inventions that might make a trade quicker.

◄ **This stained glass window** from Chartres Cathedral in France, shows carpenters at work. By the late 15th century most town craftworkers belonged to guilds.

20

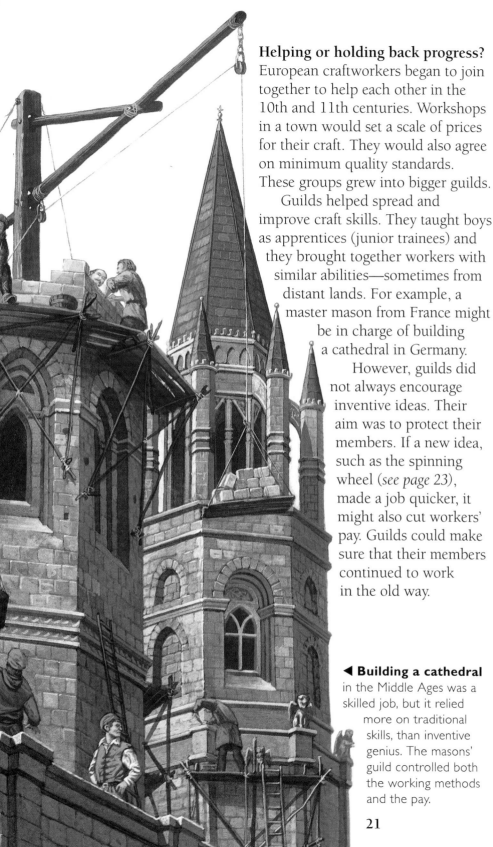

Helping or holding back progress?
European craftworkers began to join together to help each other in the 10th and 11th centuries. Workshops in a town would set a scale of prices for their craft. They would also agree on minimum quality standards. These groups grew into bigger guilds.

Guilds helped spread and improve craft skills. They taught boys as apprentices (junior trainees) and they brought together workers with similar abilities—sometimes from distant lands. For example, a master mason from France might be in charge of building a cathedral in Germany.

However, guilds did not always encourage inventive ideas. Their aim was to protect their members. If a new idea, such as the spinning wheel (*see page 23*), made a job quicker, it might also cut workers' pay. Guilds could make sure that their members continued to work in the old way.

◄ **Building a cathedral** in the Middle Ages was a skilled job, but it relied more on traditional skills, than inventive genius. The masons' guild controlled both the working methods and the pay.

DE HONNECOURT

French master mason Villard de Honnecourt (A.D. 1225–50) sketched many designs for great cathedrals and practical devices. Most of his designs illustrate architectural practices in use in the 13th century, but a few show fantasy machines. This perpetual-motion device was supposed to spin for ever without power. We now know that this is impossible.

Nameless inventors
It is not always easy to trace the origins of an invention. When the empires of Greece and Rome were powerful, famous thinkers sometimes unfairly took the credit for inventions. A spiraling water-lifting device, for example, is named after the Greek mathematician, Archimedes (287–212 B.C.). Yet he probably just wrote the first description of it. The Arab inventor of the "Archimedes screw" has long since been forgotten.

We know the names of very few inventors from the Middle Ages. It is probable that most of them were artisans who, little by little, discovered new ways of working to make their daily lives a bit easier and more comfortable.

21

The wind from the East

▲ **A rudder** guided Chinese ships from the eighth century. Four hundred years later, European mariners copied the idea when they saw rudders being used on Arab ships.

▼ **The "one-wheeled boat"** was the wheelbarrow, which only reached Europe in about A.D. 1200. The Chinese had been using them 1,000 years earlier. They added sails to pull them along.

In the Middle Ages, travelers told fantastic stories about inventions from the East which many people found hard to believe. But the tales of machines that spun thread faster than any human hand, one-wheeled "boats," and magic spoons that always pointed the same way, were all true! Chinese inventors had the answers to numerous mechanical problems long before the West. When Europeans saw the inventions with their own eyes, they overcame their disbelief and copied them eagerly. Nevertheless it took centuries for them to catch up.

▲ **This mechanical water clock**, made in China in A.D. 1088, chimed the hours. No European clock was as complex for three centuries.

THE CAMERA OBSCURA

Blocking a window with shutters darkens a room. But if there is a woodworm hole in the shutters, a tiny beam of light can enter. The beam has remarkable properties. It projects an upside-down picture of the world outside on the wall opposite. This is the principle of the camera obscura, the ancestor of the modern camera. The Chinese astronomer Shen Kua first described the camera obscura in A.D. 1086. He may have studied the effect in a specially built tower like this one.

ima
worm ho

◀ **The magnetic spoon** amused the Chinese by always pointing north. Chinese navigators later introduced the magnetic compass to Europe.

Chinese ingenuity

Why were Chinese people so much more inventive? Perhaps because of the way that Chinese society was organized. A class of officials, called mandarins, ruled China in the Middle Ages. They encouraged inventions that they believed would be useful. Europe lacked this official guidance. China's society was also more ancient, and long periods of peace had allowed its people to grow and prosper. There were other reasons as well. China has a very long coastline and great rivers. Chinese people became expert shipbuilders as they traveled on the rivers and oceans.

Silk also encouraged Chinese invention. Weavers made this luxurious fabric from the threads spun by silk moth caterpillars. Each caterpillar made a single thread nearly 1,000 yards long. Chinese inventors devised a machine to reel up the threads. This silk-reeling machine went on to become the spinning wheel.

▶ **Paper-making** was a Chinese discovery. In A.D. 751, Arab warriors took many Chinese as prisoners and stole their ideas. Europeans learned to make paper four centuries later.

Europe learns Chinese wisdom

Early travelers, such as the Venetian Marco Polo (A.D. 1254–1324), brought stories of Chinese technology straight back to Europe. However, knowledge of amazing Chinese inventions also came by more indirect routes. Some came along the Silk Road—an overland trade route linking East and West. When merchants stopped at market towns, they swapped stories. More information came by sea. Mariners brought Chinese ideas to the Arab world of northern Africa and the Middle East, and the Crusades helped spread many inventions further north to Europe.

▲ **The spinning wheel** came into use in Europe in about A.D. 1300. It enabled people to spin wool into thread far more quickly. Chinese textile workers had already been using spinning wheels for two centuries.

Sources of power

Turning in a strong breeze, the canvas sails of a windmill were a new and exciting sight in 12th-century Europe. Harnessing the wind in this way provided an extra source of power. Better still, the power was free for all to use. Mill builders of the Middle Ages also made more use of old power sources. They fitted improved wheels to water mills. These changes made it cheaper to grind grain. Growing it got easier, too. A new kind of harness enabled horses to pull heavier loads. Horse-pulled plows replaced the slower ox-drawn ones. With teams of horses, farmers cultivated more land and grew extra crops to feed hungry people.

▲ **Early European mills were post-mills.** They stood on huge wooden posts anchored upright in the ground. The miller could swivel the whole mill so the sails always faced the wind.

▼ **Windmills** were in use in Persia 200 years before they arrived in Europe. The tall, reed sails were attached to vertical wooden poles and spun (like a carousel) in the wind.

sail

pole

Wind power
The people of western Asia probably built the earliest windmills. However, when windmills first appeared in Europe some time before A.D. 1140, they looked very different. Their sails turned in an upright circle, as a wheel does, on the front of the mill. To work properly, the sails had to face the wind. The whole mill balanced on heavy posts. When the wind changed, the miller pulled the mill round to face the breeze again.

Water power
The first-century Roman engineer Vitruvius invented a new mill. It was powered by water and turned even in slow-flowing streams.

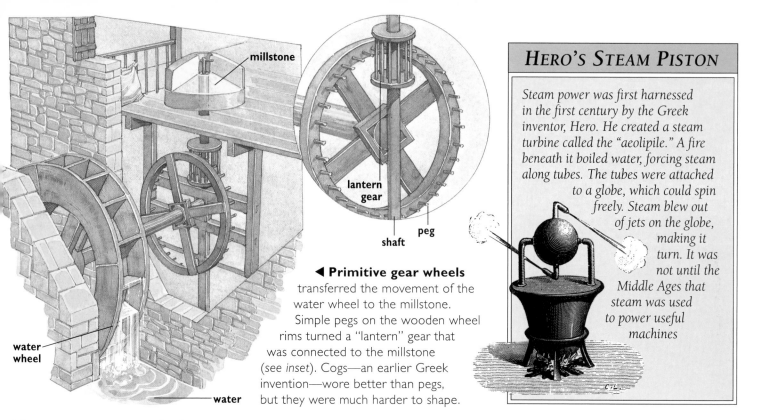

millstone

lantern gear

shaft

peg

water wheel

water

◀ **Primitive gear wheels**
transferred the movement of the
water wheel to the millstone.
Simple pegs on the wooden wheel
rims turned a "lantern" gear that
was connected to the millstone
(see *inset*). Cogs—an earlier Greek
invention—wore better than pegs,
but they were much harder to shape.

However, "Vitruvian" wheels were
not widely used until the Middle
Ages. The main use for water mills
was to grind grain into flour. But
gradually, people realized that water
power could be used in other ways.

Gears

Both wind and water mills needed
gears in order to work. Both the
windmill's sails and the water mill's
wheel stood upright. Millstones lay
flat and a simple pole could not link
them to a sail or a wheel.

However, gears could, and they
could also change the speed at
which the mills worked. Building
windmills taught engineers how to
use gears. In time, they used this
new knowledge to build countless
other ingenious machines.

Horse power

Rigid horse collars began replacing
choking straplike harnesses in 10th
century France. They greatly
increased the loads horses could
pull, and made plowing more
efficient. On the road, swift
horse-drawn carriages replaced
lumbering oxcarts.

▼ **An ox harness** allows oxen to pull
with all their strength. On a horse, a
similar harness crushes the windpipe and
stops the animal from breathing properly.
The invention of the padded horse collar
enabled a horse to pull heavier loads.

RENAISSANCE INGENUITY

A NEW AGE OF INVENTION, SCIENCE, AND GREAT ART began 500 years ago. It has been called the Renaissance—French for "rebirth"—because scholars studied works of forgotten writers and ancient wisdom was "born again."

Printing spreads knowledge

▲ **The Bible was the first book** that Gutenberg printed. He left spaces for artists to fill with colorful decorations. These made it look handwritten.

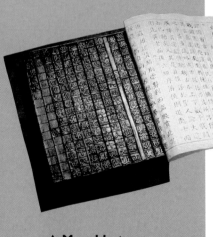

▲ **Movable type** was an 11th century Chinese invention. It was not a success because Chinese writing uses 80,000 different symbols. European printers needed far fewer pieces of type because the Roman alphabet has only 26 letters.

Books were precious objects until the 15th century. Making a book took many weeks. Monks copied every page by hand. All this changed around 1440 when a German craftworker, Johannes Gutenberg, found a way to make books more quickly. His printing process was a sensational success. Knowledge of printing spread rapidly. Soon printers were at work all over Europe. By 1500, they had produced 10 million books. With these new, cheaper books, more people learned to read and, in time, printing helped inventors with their work.

printing press

Gutenberg's big idea

Johannes Gutenberg (1400–68) did not invent printing: Chinese people printed books 500 years earlier. He *did* invent the printing press. This simple machine squeezed a piece of paper against inky type (raised metal letters). He molded the type a letter at a time by casting it from melted lead. To print a page of writing, Gutenberg arranged the letters he wanted line by line. The letters could later be separated and used again.

▼**Gutenberg's printing works** was a busy place. His employees worked 12 hours each day, printing as many as 3,000 pages on the wooden press.

lead type

LEONARDO DA VINCI

Leonardo da Vinci (1452–1519) lived in Italy. He was a brilliant painter, a daring inventor, and a great scientist. He filled sketchbooks with clever ideas—such as this helicopter—that looked forward to the 20th century. Nowadays, he represents the joining of art and science, which was a leading Renaissance ideal.

▶ **Oil paint** enabled Renaissance artists such as Jan van Eyck (c.1390–1441) to record very fine detail as in this painting. The inks that Gutenberg used were similar to oil paint.

Ink and paper

Gutenberg's press and movable type would not have worked without two other inventions. The first was a new kind of ink. Writing ink was made from water and did not stick to the metal letters. Instead, Gutenberg used oily ink that stuck to the type and copied the letter shapes clearly.

The second crucial invention was papermaking. Paper was still a new material when Gutenberg began printing. Indeed, he printed between 35 and 50 Bibles on vellum (calf skin). Paper was much cheaper, and

Gutenberg printed four times more paper Bibles than vellum Bibles.

Gutenberg helped make books available to everybody, but printing didn't make him rich, as he was a bad businessman. His partner, Johann Fust, later made printing a financial success.

Measurement and observation

In Middleburg, Holland, an optician's apprentice played with two spectacle lenses. He held one to his eye and the other at arm's length. When he looked through both lenses at a distant church, he was surprised to see that it appeared to be much closer. The boy (whose name has long since been forgotten) had unwittingly invented the first telescope. His master, Hans Lippershey (c.1570–1619) instantly realized the importance of the discovery and developed the idea further.

The invention of the telescope, in 1608, was of tremendous value to science. The thermometer, clock, and microscope were also perfected around the same time. They all greatly improved measurement and observation and gave people a better understanding of the world.

pendulum

▲ Galileo's clock kept time with a pendulum, but only if you pushed it with your finger! This reconstruction is based on drawings made by his son or a friend.

▲ The reflecting telescope, devised in 1670 by Isaac Newton (1642–1727), used a dish-shaped mirror for clearer views.

▼ John Kepler's telescope improved on Galileo's instrument. The German astronomer (1571–1631) altered the lenses so that it could magnify up to 1,000 times.

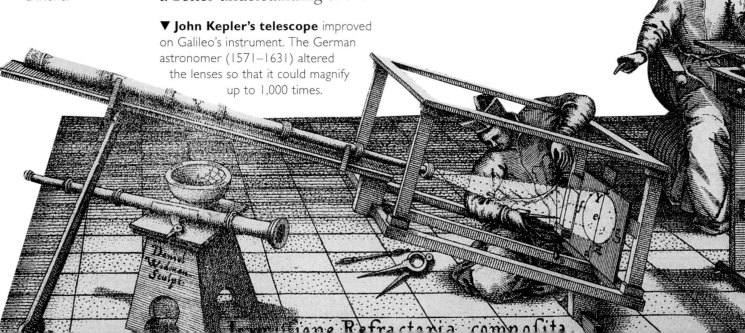

The telescope

Hans Lippershey at first thought his "instrument for seeing at a distance" would be valuable in warfare so he tried to keep it secret. But news of it quickly spread and within a year the Italian scientist Galileo (1564–1642) heard of the device and constructed a telescope which could magnify even more—up to 30 times.

Time and temperature

Galileo played an important part in the invention of the clock and the thermometer. While in Pisa cathedral, he noticed that however far a lamp swung, each swing took the same time. Many years later this led him to plan a pendulum clock, but he died before he had time to complete it.

Galileo also experimented with measuring temperature. He poured water into a long tube with a bulb at the top, and dipped the other end of the tube into a beaker of water. Warming the bulb with his hand made the air inside expand, which forced the water down the tube. The resulting rising water level in the beaker was a crude measure of the temperature of his body.

▶ **The thermometer**
led to many important discoveries about the human body, and about climate. This example was made by Galileo in Italy around 1592.

MICROSCOPES

Magnifying glasses were in use in the first century, but they enlarged images of objects only a few times. By the 1670s, the Dutch cloth merchant Anton van Leeuwenhoek (1632–1723) had vastly improved them. He used tiny lenses to make microscopes that magnified objects 300 times. Modern microscopes are only four times more powerful.

Using two magnifying glasses together increases their power. A pair of Middleburg opticians, Hans Jansen and his son Zacharias, made this discovery in 1590. The device they invented is called the compound microscope. Improvements in the 17th and 18th centuries turned it into the instrument that scientists use today.

▼ **Leeuwenhoek's microscope**
was small enough to fit in the palm of his hand. Its single lens was the size of a raindrop.

▶ **Campani's microscope**
was the first to have a threaded barrel. Turning it focused the image. Guiseppe Campani (1635–1715) was one of Italy's greatest instrument-makers.

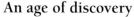

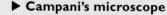

An age of discovery

Scientists were quick to use these new instruments. When Galileo looked at the planet Venus with his telescope, he was able to see that it had phases like the Moon. Although the phases of Venus lasted a lot longer than the phases of the Moon, its appearance changed from a disk to a crescent as it circled the Sun and the Sun's light fell on different parts of its surface.

This was an astonishing discovery as it proved that the Earth moved round the Sun. The Catholic Church mistakenly believed that the Earth (not the Sun) was the center of the universe and imprisoned Galileo because he disagreed with them. While he was in prison, he studied the pendulum clock.

A century after the invention of the pendulum clock, Venus passed between the Earth and the Sun. Scientists used clocks to time this "transit" at different places on the Earth's surface. They used the results to work out the distance from the Earth to the Sun.

The microscope also led to astonishing new discoveries. With it, Leeuwenhoek was able to see and sketch bacteria and other minute forms of life for the first time.

Navigation and new weapons

Early European maps of the world look oddly incomplete. There are gaps where North and South America should be. Europeans knew almost nothing of America until the end of the 15th century, when mariners crossed the Atlantic. Several important inventions made their voyages possible. The compass and other instruments helped them find their way and improved ships made sailing safer. When Europeans reached distant shores, they claimed the lands they found as their own. Native people fought these foreign invasions, but were no match against the new weapons, guns, and explosives.

cross-staff

magnetic compass

◄ **Using an astrolabe**, sailors could measure the height of the Sun at noon. This told them how far north or south they had sailed.

Ships and navigation

The Spanish began to make ships strong enough for long voyages at the end of the 15th century. They started with a rigid skeleton and nailed planks to it. With extra masts and sails, the ships could cross the Atlantic in about a month.

The first mariners that sailed these ships steered using a compass. Its balanced magnetic needle always pointed north. Mediterranean sailors learned to use this Chinese invention (*see page 23*) in the 12th century. Arab navigators probably taught them. They may also have given European mariners instruments to measure latitude—the distance they had sailed to the north or south of the equator.

◄ **The mariner's astrolabe** was in use from the 13th century. It was a much simpler version of an earlier Arabic instrument.

1482

1607

 ◄ Navigational instruments such as these helped mariners discover new lands. The star-shaped compass pointed the way. The 16th-century cross-staff showed how far north or south their ships had sailed. The quadrant, invented in 1731, measured latitude more accurately than before.

quadrant

Improving maps

Mariners began to draw their own charts (sea maps) in the 13th century. Mapping the world, though, was more difficult because the Earth is round and a map is flat.

Flemish cartographer (map-maker) Gerardus Mercator (1512–94) solved the problem. His 1569 world map flattened the globe by "stretching" it into a tube. This map was the first to show mariners the correct course to steer. But only places on the equator appeared the correct distance apart.

Weapons

Chinese alchemists (mystical scientists) invented gunpowder in the ninth century. They made it into fireworks. Foreign warriors realized that the explosive powder could be a useful weapon. Europeans learned about gunpowder before 1280. They used it to fire cannonballs.

Hand weapons were a later invention. At first they looked like small cannons on sticks. But by the mid-15th century, Spanish soldiers had begun to fight with the harquebus. This was the first gun that soldiers could fire from the shoulder. There was no trigger. The soldier lit a smoldering fuse. Lowering it into a pan of powder fired the gun.

Within a century inventors had devised guns that were easier to use. Pulling the trigger of a flintlock made a piece of flint strike a steel plate; this caused sparks which made the gun fire.

▲ Maps of the Earth looked very different before (as in this map of 1482) and after the discovery of the "New World" (as in 1607).

TRAVELERS

European mariners were not the only travelers making spectacular journeys. Chinese admiral, Cheng Ho (1371–1435) made seven ocean voyages. He sailed as far as the African coast. Arab travelers also wandered far afield (left). Ibn Battutah (1304–68) spent 30 years exploring as far as Beijing and Siberia. If the roads he traveled were stretched out straight, they would go eight times around the world!

▲ Cannons destroyed strong castle walls and changed the way armies fought wars. Soldiers armed with a cannon could kill advancing archers long before they were within range of their arrows.

AN INVENTIVE AGE

IN THE 18TH CENTURY, A SERIES OF DRAMATIC INVENTIONS TRANSFORMED daily life. Factories equipped with iron machines replaced small workshops with skilled workers. This time of change is called the Industrial Revolution.

New materials

When ironmakers fueled their furnaces with charcoal (partly burnt wood) 300 years ago, iron was a scarce and valuable metal. To produce it they cut down Europe's great forests, and shortage of wood kept the price of iron high. But in 1709 an Englishman discovered how to make iron using coke (roasted coal). Coal was cheap, so iron production grew quickly, and iron soon replaced the weaker wooden parts in factory machines. The invention of underwater-setting cement and steel in the 19th century provided builders with further scope for development.

▼ **Glass blowing** at a glass furnace. One of the many skills introduced from France and Italy in the 13th-16th centuries.

Imported knowledge
Ironworking, steam power, and fabric mills helped make England a leader of the Industrial Revolution. But in earlier centuries Germany, France, and the Netherlands were more advanced. England caught up by attracting craftworkers and engineers from abroad. When nobody in England could extract copper, Queen Elizabeth I (1533–1603), called for the expert help of German engineers, and in 1587 the English government encouraged Flemish wool weavers to move to England. The following century, cotton weavers and dyers came from the Netherlands and Dutch engineers drained English marshes.

▲ **A water pump** was one of 200 machines that the Italian engineer Agostino Ramelli (1531–90) sketched in 1588. Books of inventive sketches such as this were known as "theaters of machines."

32

► The Eddystone lighthouse, off the southern coast of England, got its strength from hydraulic (underwater-setting) cement. In 1756, the architect John Smeaton was the first to study cements scientifically.

stones interlocked to resist pounding waves

mortar joints made from hydraulic cement

◄ Dyers from the Netherlands taught English clothworkers how to color wool in the 17th century. Up until then, English weavers sent cloth abroad to be dyed.

▼ Silk weavers from the French town of Rouen established the first silk looms, such as this one, in England in 1532.

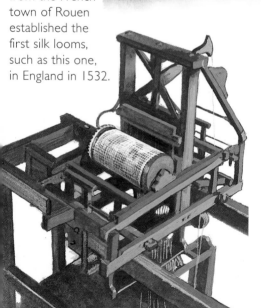

Crucial metals

Abraham Darby (*see right*) provided the huge iron cylinders that Thomas Newcomen needed to build his steam pumping engines (*see page 34*). These engines pumped water from mines, which made coke production cheaper and, in so doing, reduced the cost of iron. Cheap iron reduced the cost of Newcomen engines. With this mutually beneficial relationship, industrialization moved forward in leaps and bounds.

Wrought iron and steel

The carbon in cast iron made it weak when it was stretched. However, in 1784 Englishman Henry Cort (1740–1800) found a way of removing the carbon to make wrought iron. In 1856 Henry Bessemer (1813–98) discovered a way to remove even more carbon to make the stronger metal, steel.

THE DARBY FAMILY

At his works in Coalbrookdale, northwestern England, Abraham Darby (1750–91) first discovered how to make iron using coke. His son and grandson, who were also called Abraham, helped to make the metal cheaper and more widely available. Abraham II made parts for the first steam locomotive (see page 35); Abraham III made the first iron bridge (see below).

The furnace at Coalbrookdale

▼ This cast-iron bridge, which still stands in Ironbridge Gorge, England, was one of the "Wonders of the Modern World" when it was completed in 1779.

New power and locomotion

"Rail travel at high speed is not possible," wrote a university professor around 1830, "Passengers, unable to breathe, would die." But within 10 years locomotives were traveling at 55 mph and not one passenger suffocated! Railways made travel fast and cheap, and lines grew quickly. By 1860 there were about 120,000 miles of track worldwide.

Railway trains were not the first use of steam power, though. The steam age began in 1712, with slowly rocking pumps that lifted water from mines. When they were improved to turn wheels, steam engines replaced water power in mills all over England.

▶ **Newcomen's pump**
sold well. It made deeper mines possible, increasing production of coal and metal.

cylinder

rod powered pump at mine bottom

Newcomen's engine

The English blacksmith Thomas Newcomen (1663–1729) built the first steam pump to empty flooded mines. Steam filled a huge cylinder, forcing up a piston inside. Cooling the cylinder condensed the steam (turned it to water). This took up much less room, so the piston was sucked in. By repeating these actions Newcomen set a wooden beam rocking. The beam operated a pump which lifted about 60 buckets of water a minute.

Newcomen's engines began working in 1712. They were slow and costly but worked without rest, unlike the 11 horses each replaced.

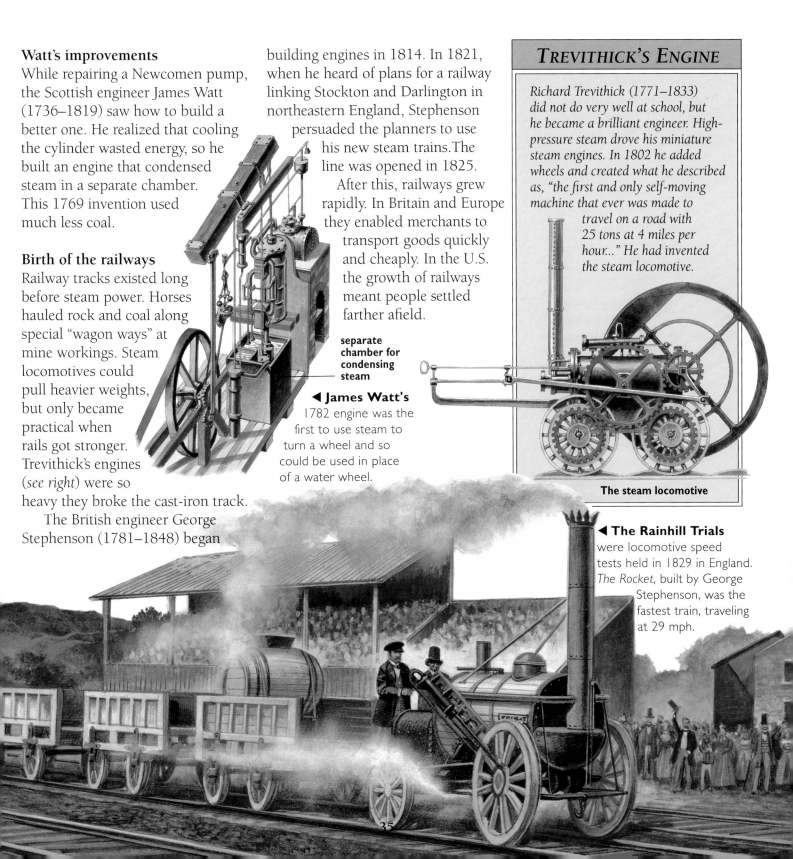

Watt's improvements

While repairing a Newcomen pump, the Scottish engineer James Watt (1736–1819) saw how to build a better one. He realized that cooling the cylinder wasted energy, so he built an engine that condensed steam in a separate chamber. This 1769 invention used much less coal.

Birth of the railways

Railway tracks existed long before steam power. Horses hauled rock and coal along special "wagon ways" at mine workings. Steam locomotives could pull heavier weights, but only became practical when rails got stronger. Trevithick's engines (*see right*) were so heavy they broke the cast-iron track.

The British engineer George Stephenson (1781–1848) began building engines in 1814. In 1821, when he heard of plans for a railway linking Stockton and Darlington in northeastern England, Stephenson persuaded the planners to use his new steam trains. The line was opened in 1825.

After this, railways grew rapidly. In Britain and Europe they enabled merchants to transport goods quickly and cheaply. In the U.S. the growth of railways meant people settled farther afield.

separate chamber for condensing steam

◄ **James Watt's** 1782 engine was the first to use steam to turn a wheel and so could be used in place of a water wheel.

TREVITHICK'S ENGINE

Richard Trevithick (1771–1833) did not do very well at school, but he became a brilliant engineer. High-pressure steam drove his miniature steam engines. In 1802 he added wheels and created what he described as, "the first and only self-moving machine that ever was made to travel on a road with 25 tons at 4 miles per hour..." He had invented the steam locomotive.

The steam locomotive

◄ **The Rainhill Trials** were locomotive speed tests held in 1829 in England. *The Rocket*, built by George Stephenson, was the fastest train, traveling at 29 mph.

New methods and machinery

Before the Industrial Revolution everything was made by hand by skilled craftspeople in small workshops. But when they could not produce enough, people built machines to speed up production.

The English fabric industry was the first target. Inventors devised ingenious machines for spinning and weaving. Powered by water wheels, they were far quicker and unskilled workers could operate them. These machines briefly gave English industry an advantage, as laws prevented the export of the new machines. However, the mechanics who operated them emigrated to Britain's American colonies. There they built new machines—and improved them. Soon industry was growing faster in America than in England.

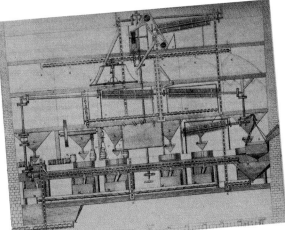

▲ **Arkwright's spinning frame** spun cotton fibers into yarn four bobbins at a time.

bobbin

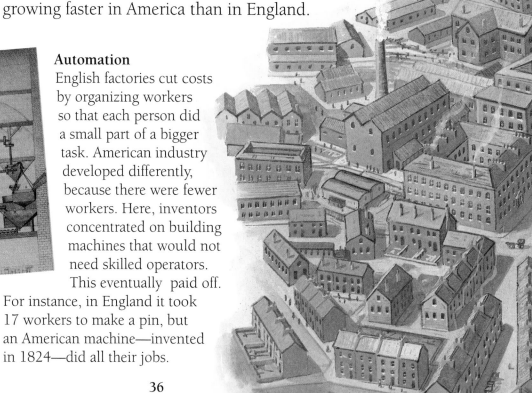

▲ **The designs that Oliver Evans** (1755–1819) drew of his flour mill (1784). It was the first completely automated factory.

Automation

English factories cut costs by organizing workers so that each person did a small part of a bigger task. American industry developed differently, because there were fewer workers. Here, inventors concentrated on building machines that would not need skilled operators. This eventually paid off. For instance, in England it took 17 workers to make a pin, but an American machine—invented in 1824—did all their jobs.

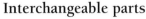

Factories

Spinning and weaving machines were built beside rivers as they needed water to turn the wheels that powered them. English inventor Richard Arkwright (1732–92) was the first to use waterpower with machine production. In 1771, he installed spinning machines in a mill in Cromford, England.

Factories like this could make cheaper goods than craft workshops. They expanded quickly, and as workers moved from the countryside to work in mills, towns grew up around them.

Interchangeable parts

If part of an 18th-century musket broke, the gun was useless. Every musket was made by hand, so each one was slightly different.

The idea of making parts interchangeable (able to replace each other) began in France in the 1780s. However, the process was difficult and expensive. American Simeon North (1765–1852) was among the first to produce muskets by this method, about half a century later.

As factories made more interchangeable parts, the cost of their products fell. Automatic machines and better organization of work cut costs even further. This combination of improvements became known as "the American system of manufacture." Using this system, American factories could make goods more cheaply and in greater numbers than any other country.

◀ **Industrial towns** grew up around busy factories. Canals and railways brought in raw materials, and carried away finished products. The towns attracted poor people eager for work and many lived in crowded, unhealthy slums dominated by smoking factory chimneys.

▲ **American muskets** were mass produced in factories using new machines and new ways of management. Their interchangeable parts made them more costly to produce than those made in the traditional way. However, being easier to repair, they were cheaper in the long term.

WORKING LIFE

Not everyone shared the benefits of the Industrial Revolution. Factory owners became wealthy, but those who operated the machines worked very long hours for low pay—like the workers in the cotton mill below. Factories in England employed children as young as seven. Some worked 15 hours every day. Others worked in darkness down mines.

Conditions for adults were no better. Manufactured goods became cheaper—but factory workers could not afford to buy what they made.

▲ **Thomas Edison**
(1847–1931) is one of the
United States' most brilliant
inventors. He devised not
only the light bulb, but also
the gramophone and
the microphone.

THE ERA OF ELECTRICITY

WE TAKE IT FOR GRANTED THAT NIGHT TURNS TO DAY when we flick a switch, yet the light bulb is little over a century old. Its invention signaled the coming of a new age of technology that transformed transport, communications, and entertainment.

The mighty light bulb

▶ **Edison used** a bamboo filament in his first light bulb. The lack of air in the bulb, prevented the bamboo burning up.

The first electric lights were a little too successful. They were so bright that they were useful only in lighthouses and streets. Inventors struggled to "subdivide" them—to make lamps glow just brightly enough to light a room. It was a long search, but one American inventor, Thomas Edison, did not give up easily. "I've tried everything," he said. "I have not failed. I've just found 10,000 ways that won't work." In 1881 he finally succeeded. English inventor Joseph Swan devised a similar lamp around the same time.

◀ **Gas lamps** lit streets from about 1800. Indoor gas lighting soon followed. Gas lighting had no real competitors until the invention of electric lights.

▶ Arc lamps, invented in 1847, created a brilliant light from an electric spark. They were so bright they could cause blindness.

The search for a filament
To make light bulbs, Edison and Swan enclosed filaments (thin strands) of metal wire in glass bulbs and sucked out the air. Passing a current through the filaments made them glow, but melted the wire. The two inventors made long-lasting bulbs only when they replaced metal filaments with carbon filaments.

▼ Alessandro Volta (1745–1827) demonstrated his battery to the French Emperor Napoleon in 1802, but electricity had few uses until the invention of the generator.

The slow advance
It's hard to imagine life without electricity now, but its advantages were not obvious in 1880 when it could only be used for lighting, and oil or gas lights were cheaper. People were also afraid of getting electric shocks. When electricity was first installed in the White House in 1889, the president would not touch the switches!

The electric power industry grew slowly because there was little need for electricity by day. This changed when buses began using electric power around 1890.

Industrial use also grew. Homes were especially slow to take up the new source of power. By 1921 only one in eight British homes had electricity.

Supplying electricity
On its own a light bulb is useless. It needs a constant supply of electrical power. Until 1832 batteries were the only source of electricity. But then the English scientist Michael Faraday (1791–1867) found that a moving magnet made electricity flow in a coil of wire. This discovery led to the generator, or dynamo. Generators were too costly for every home to have one. Instead, central generating stations produced the power, which was supplied to homes and businesses by cables.

◀ Irons were among the first household electrical goods. People screwed them into light sockets to make them work.

THE GRAMME DYNAMO

Driven by a steam engine or a water wheel, dynamos turned movement into electricity. Early dynamos were vast and inefficient as they did not produce a continuous current, and some got so hot that they had to be cooled with water. In 1870 the Belgian electrical engineer Zénobe Théophile Gramme (1826–1901) devised a dynamo that avoided these problems.

▶ Gramme's dynamo helped the electrical industry grow in France where he made it.

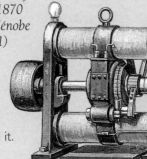

the Rover
1885

Transport on land and sea

▲ Bicycles were invented in 1839 by a Scottish blacksmith, Kilpatrick Macmillan (1813–1878). Modern "safety" bicycles, such as the Rover, appeared 50 years later. They had chains, spokes, and inflatable tires.

When a steamship first crossed the Atlantic Ocean in 1819, its arrival in Ireland caused much alarm. Smoke from the boilers was mistaken for a fire on board! By 1830, however, steamships were built from iron instead of wood and were a common sight. Together, iron hulls and steam power made oceangoing ships bigger and faster. New sources of power also brought advancement to road transport. By the end of the 19th century petrol-powered motor cars were replacing horse transport, and the bicycle was widely used.

▼ Early submarines, like this 1776 *Turtle*, are now hard to take seriously. They were not practical until 1900 when Irish-American John Holland (1840–1914) added electric motors and periscopes, transforming them into effective sea vessels.

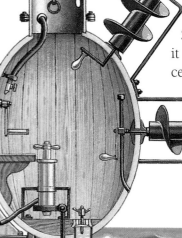

Crossing the oceans

When the 19th century began, all large vessels were sailing ships made from wood. By the end of the century, wooden hulls had given way to iron, and steam power had replaced sails. The American inventor Robert Fulton built the first successful steamer in 1807. Splashing paddle wheels drove it forward. By the middle of the century screw propellers had replaced paddle wheels, and they were as fast as sailing ships.

Shipbuilding in iron began around 1820. As iron is a stronger material than wood, these new ships could be much bigger. By 1853 engineers were building vessels that were twice as big as earlier wooden ones.

◄ **Inflatable tires** were invented twice. The first time, in 1845, they were fitted to horsedrawn carriages and attracted little interest. But in 1889, John Dunlop (1840–1921) made them a huge success when he fitted them to bicycles. Inflated tires, like the ones in the photograph, later proved ideal for use in cars.

▲ **The Benz** motorized "horseless carriage" (1885) traveled very slowly.

The internal combustion engine

Frenchman Étienne Lenoir (1822–1900) was the first to build an engine inside out! In a steam engine, combustion (burning) on the outside turns water to steam, and the steam drives a piston. Lenoir's engine got rid of water altogether. Burning gas inside it moved the piston, so it was called an internal combustion engine.

Half a century later, the German engineer, Gottlieb Daimler (1834–1900) built a gasoline engine. It was used to power the first cars.

The horseless carriage

Horses had no competitors in the transport business until the 1880s. Electricity and steam powered a few vehicles, but their engines were heavy and slow, or quickly ran out of fuel.

Internal combustion engines had none of these disadvantages, and several engineers used them to build "horseless carriages." German Karl Friedrich Benz (1844–1929) was one of the first to do so. In 1885 he drove his motorized tricycle at about the speed of a trotting horse. Four-wheeled vehicles evolved a few years later.

◄ **The Great Eastern,** designed by English engineer Isambard Kingdom Brunel (1806–59), had masts and sails as well as paddle wheels and a propeller. It was as long as nine tennis courts, and was the world's biggest ship for 40 years.

AGE OF THE ROAD

Unpaved roads were dusty tracks in summer and swamps in winter, and their poor condition slowed transport. Scotsman John McAdam (1756–1836) realized that a road would not break up if water were drained off it. "Macadamized" roads made travel faster and safer from 1815, and the popularity of the bicycle led to further improvements at the end of the century. The bicycle thus "paved the way" for motor travel.

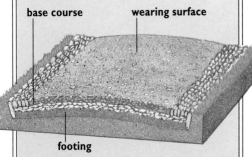

base course   wearing surface

footing

▲ **McAdam's roads** were curved so that rain-water, which might otherwise have weakened the foundations, drained off them.

Discovering flight

Inventors have dreamed of flying like birds since the days of ancient Greece. Some even strapped on wings and jumped from cliffs flapping their arms! But flying machines made progress only when inventors ignored the flapping of a bird's wings, and copied the gliding flight of larger birds. The albatross keeps its outstretched wings still and the draught of air across its curved feathers provides the lift it needs to fly. The English engineer George Cayley (1773–1857) was the first to copy this when he experimented with fixed-wing craft, and in 1853, he built a glider which could carry a pilot.

German Otto Lilienthal (1848–1896) had even more success. He made hundreds of flights, soaring to 1150 feet. He inspired other experimenters, including the American Wright brothers.

▲ **Balloons first flew** in 1783. The French Montgolfier brothers filled a huge silk balloon with warm air that lifted it from the ground. Hydrogen balloons flew the same year.

The Wright brothers

Wilbur (1867–1912) and Orville (1871–1948) Wright had experimented with kites as boys, and reading about Lilienthal's gliders reawakened their interest in flight. Scientists believed flight was impossible, but the brothers realized that to build a flying machine, they would need to solve three problems: lift, control, and power.

They tackled lift first. In 1900 they traveled to Kill Devil Hills on the windy North Carolina shore. There they flew a kite big enough to lift one of them. Wilbur soon worked out how to control the kite's direction of flight by twisting the wings. The brothers returned to the beach in 1901 and 1902, taking an improved glider each year.

Wilbur Wright

Orville Wright

In 1903 they invented the propeller, which provided power. Orville's historic first flight—on December 17th 1903— lasted less than a minute, but was a triumph.

▲ **Otto Lilienthal** invented a kind of hang glider. He built it from canvas stretched over a willow frame.

AIRSHIPS

The first balloons drifted with the wind; their passengers could not choose their destination. This changed in 1852 when Henri Giffard (1825–1882) built an airship—a powered balloon. An 11-foot propeller drove it along, and a rudder steered the craft. Airships provided the first transatlantic passenger service.

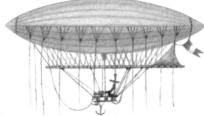

▲ **Airships** used engines to fly against the wind.

▼ **The Hindenburg airship** caught fire while landing in 1937, killing 35 people.

◄ **Flyer I was the Wright brothers'** famous 1903 aircraft. Its engine was tiny— no bigger than that of a modern motor cycle.

Air travel takes off

The value of flying machines in warfare was recognized from the first balloon flights, and the American Army took a keen interest in the Wrights' work. By the time World War I (1914–1918) began, all the major nations of Europe had flying units. Aircraft have played a major role in every war since.

Commercial airlines began straight after the war, and today, flying is the most practical method of long-distance travel.

The rise to fame

The Wright brothers' achievement surprised the world (and their father!), and made them famous.

They continued to improve their aircraft, and Flyer III, built in 1905, could stay in the air for about half an hour. It could also fly in circles and figure eights. By 1908 the Wright brothers were giving public flight demonstrations.

▶ **The first helicopter** built by Paul Cornu (1881–1944) left the ground four years after the Wrights' first powered flights. It flew for just seconds and rose only to head height. Practical helicopters did not fly until 1936.

Communications and entertainment

Pony Express was the fastest way to send messages across the United States in 1860. Riders such as Buffalo Bill galloped the 1,800 miles in just 10 days! However, the service lasted only 18 months, as it could not compete with the new telegraph wires that could transmit a message in minutes.

Fifteen years later, speech crackled along the wires, and by 1895 radio made wires unnecessary. Photography, then 20 years old, recorded the brief life of Pony Express. But movie cameras were invented 30 years too late to film the ponies.

peep hole

cogs driven by simple motor

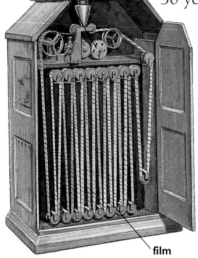

film

▲ **Thomas Edison's kinetoscope** showed a jerky movie for a minute. One viewer at a time peeped through a hole at the top to see the film.

▶ **Praxinoscopes** and other optical novelties were popular shortly before the invention of the kinetoscope. They made drawings or photographs move by showing each of them briefly.

Cameras and movies

The French chemist Joseph Niepce (1765–1833) experimented with photography in 1826. He fixed a lens to a small box. The lens projected an image on to a tar-coated metal plate. The tar hardened where light struck it. Cleaning soft tar away created the world's first photograph.

handle attached to picture disk

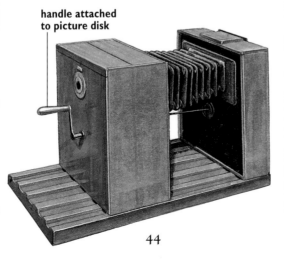

▲ **Fox Talbot's window,** which he photographed in 1835, was the world's first negative. He used a very simple wooden box camera. Talbot's process later evolved into modern photography.

In 1839, Niepce's partner, painter Louis Daguerre (1787–1851) perfected the "daguerreotype," using a copper plate coated with silver iodide. Around this time English physicist, William Fox Talbot (1800–77) also invented a photographic process.

However, it was another Englishman William Dickson (1860–1935), who made some of the first moving pictures. He worked for Thomas Edison (*see page 38*) and built a camera that took 46 photographs each second on a roll of film. Viewing the processed film strip in a kinetoscope gave the flickering illusion of movement.

► **Crystal sets** first brought radio into the home. To tune in to a station, listeners used a wire to probe a crystal in the set.

Hemmets Journal

▼ **Radio sets** became popular at home once voice broadcasting started in the 1920s.

MORSE AND TELEGRAPHY

Telegraphy means "distant writing" in Greek. A British railway company installed the first electric telegraph in 1837. But it was an American artist, Samuel Morse (1791–1872), who made it a success. With Joseph Henry (1797–1878) and Alfred Vail (1807–59), Morse devised a switch that sent short pulses of current (dots) and long ones (dashes) along a wire. A receiver at the other end marked the dots and dashes on a strip of paper. Morse devised a code of dots and dashes to represent letters and numbers. The telegraph caused a sensation when the first 40-mile line opened in 1844.

▼ **Morse's telegraph receiver** of 1844 marked dots and dashes on paper. Telegraph operators quickly learned to code messages.

Wires and wireless

Messages can only be sent short distances with coded flags or fire beacons. With electricity, only the length of the wire limits how far the signal travels. The electric telegraph (*see above*) was not Samuel Morse's idea. But the system he invented got the world "wired up." At first the wires buzzed with Morse code, but in 1876, Scots-born American Alexander Graham Bell (1847–1922) discovered how to send speech down the wires. Though they look very different, today's telephones work in much the same way as Bell's invention.

▶ **Alexander Graham Bell** (*front*) invented the telephone by accident when looking for a way to send several telegraph messages along a single cable.

The Italian Guglielmo Marconi (1874–1937) got rid of the wires. His 1895 invention broadcast only Morse code, so he called it wireless telegraphy. Today we call it radio. In 1901 he broadcast the first wireless messages across the Atlantic—a new era of communication had begun.

► **Marconi's wireless** used a Morse key which was connected to a spark discharger. When the Morse key was pressed the discharger generated radio waves that could be read by a receiver some distance away. This meant that messages could be sent without wires.

TEAMWORK

As technology developed, lone inventors teamed up with scientists, and started the first research and development laboratories.

Menlo Park

Thomas Edison invented many things, but perhaps his most important contribution to technology was not an invention at all. It was a whole new way of inventing. In 1876 Edison moved to Menlo Park, near Newark, New Jersey. There he built what he called an "inventions factory." Today we would call it a research and development laboratory. Edison's aim was to succeed at invention not through luck but through careful scientific study.

▲ **Edison's "speaking" phonograph** recorded sound as a wiggling spiral groove on a tinfoil cylinder.

▶ **Menlo Park** was built in the countryside 12 miles outside Newark. Edison's father Sam chose the site and organized the building of the laboratory.

▲ **Thomas Edison** *(far right)* testing a phonograph in 1900. Edison did not always share the credit for his inventions.

▶ **The Menlo Park team** had to be as energetic as their leader. Edison *(far right)* joked that people applying for jobs wanted to know only two things, "how much we pay, and how long we work." His reply was invariably, "Well, we don't pay anything, and we work all the time."

A team of experts

When setting up his Menlo Park laboratory, Edison recognized that he could not do everything himself if he were to achieve his aims. For instance, he did not pretend to be good at mathematics: "I do not depend on figures at all," he wrote. "I try an experiment and reason out the result, somehow, by methods which I could not explain."

To make up for this, Edison hired the mathematician and physicist Francis Upton (1852–1921). Scientists with knowledge of other special areas joined Upton at the lab. Edison worked at Menlo Park for six years. His team made some great original discoveries, such as the carbon microphone and the electron tube. The search for a light-bulb filament also took place there. Much of the work, though, was aimed at developing and perfecting existing inventions such as the telegraph.

Edison's imitators

Other companies were slow to learn from the example of Menlo Park. General Electric built laboratories in 1900, and Du Pont and Kodak a few years later. Today, though, research and development is an important part of every industry.

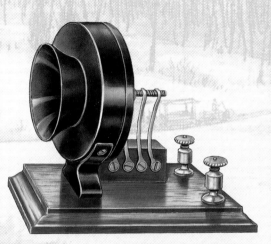

▲ **The carbon microphone,** once part of every telephone, was one of Edison's inventions.

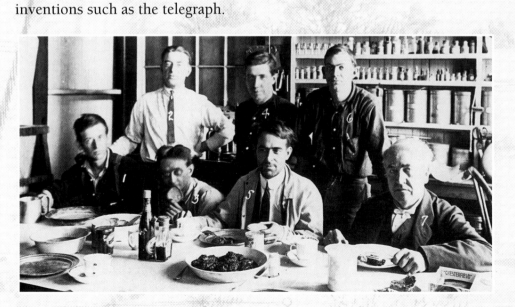

Team discoveries

Two American schoolboys dreamed up color film as they played duets on violin and piano and 20 years later, with the help of a team of scientists, their idea became a reality.

The transistor, too, was a group effort. Its three inventors won fame and a Nobel prize for their work at Bell Laboratories in New Jersey. More research and experiment followed before transistors were cheap and reliable.

However, not every great idea needs development in a big, well-funded laboratory. A pair of inventors working in a garage devised a computer that changed a whole industry.

The quest for color

Music drew together teenagers Leopold Godowsky (1899–1983) and Leopold Mannes (1899–1964). But both were also keen amateur photographers, and they were unhappy with the complex cameras needed to take color pictures. They decided to invent something better. Their experiments began at school, and in 1917 their parents lent them $800 so the two friends could continue their experiments at home.

By 1922 they had made crude color films, and the camera company Eastman Kodak agreed to help them perfect their process.

Nine years later they moved into Kodak's research laboratories where they were nicknamed "God" and "Man." Lesley Booker made dyes that added the colors to their films, Clyde Carlton coated the strips of film, and Ronald Scott processed them. It took "God" and "Man" four more years to make color film that worked.

▲ **Godowsky and Mannes** timed processing steps by humming pieces of their favorite music.

▲ **The first color film** that could be used in an ordinary camera produced small "slides" (pictures on clear film). A projector enlarged them for viewing. Since color film's introduction millions of rolls have been sold in every country of the world.

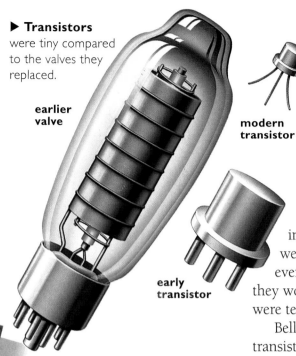

► Transistors were tiny compared to the valves they replaced.

earlier valve

modern transistor

early transistor

Inventing the transistor

In your home, millions of transistors keep everything from toasters and refrigerators to computers working. Miniaturized on computer chips, transistors switch electrical currents on and off.

In the 1950s transistors replaced valves, which looked like small light bulbs. Valves were as fragile as bulbs and got just as hot. The transistor was invented by John Bardeen (1908–91), Walter Brattain (1902–87), and William Shockley (1910–89) at the Bell Laboratories, one of the largest and best research organizations in the world today.

Manufacturing

Making transistors work in an experiment was a great achievement. But manufacturing them turned out to be very difficult. (This is true of many inventions.) Four out of five transistors did not work at all. Of the remainder, no two worked in quite the same way and all were unreliable. One company even hit the transistors until they worked. Those that did work were tested, and fitted with wires.

Bell scientists continued to lead transistor research in the 1950s. But for a fee, the company shared its knowledge with others. In 1951 four American companies were making transistors; five years later there were 26 companies in operation. Teams of scientists then competed to make transistors cheap and more reliable.

◄ Spreadsheets were invented in 1979 by Americans Dan Bricklin and Bob Frankston. These computerized accounting programs are more foolproof and much quicker than previous accounting systems.

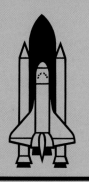

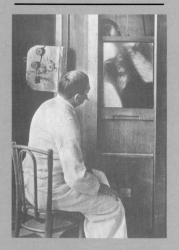

▲ **Wilhelm Röntgen**
discovered that X-rays from
a vacuum tube pass through
flesh so bones can be seen
as shadows on a
glowing screen.

INVENTION AND THE MODERN WORLD

TODAY INVENTIONS SURROUND US. THEY HELP US GROW FOOD, PROTECT OUR health, build houses, travel, and communicate. A few are completely new ideas, but many more are older inventions combined in new ways.

Medicine

Medical researchers are not usually called "inventors."
However, their work has changed our lives dramatically.
Children born today will live about 25 years longer than
those born in 1900, and their lives will be healthier.
Doctors have new tools to find out why we are sick, and
new drugs to make us well again. Now, they can also
replace worn body parts with artificial parts.

▶ **"Spare parts"** such as
artificial limbs have helped
wounded soldiers since
ancient times, but
implanting parts within
the body was rare until
the invention of new
plastic materials and
drugs in the 1950s.

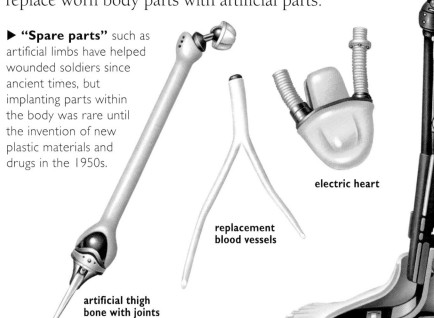

electric heart

**replacement
blood vessels**

**artificial leg
with thigh,
shin, and foot**

**artificial thigh
bone with joints**

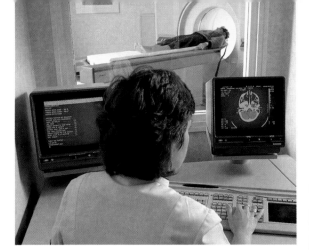

▶ **Medical scanners** show the body's internal structure on a computer monitor, so the patient does not need to be cut open.

Drug research

Modern drug research began in the 19th century when scientists purified the chemicals found in existing plant cures. Polish scientist Paul Ehrlich (1854–1915) was among the first to make drugs without plants. In 1910 he discovered the drug Salvarsan. He used it to treat syphilis, a sexually transmitted disease. Penicillin, the first antibiotic, was discovered in 1928 by Scottish bacteriologist Alexander Fleming (1881–1955).

Diagnostic imaging

The invention of X-rays in 1895 by German physicist, Wilhelm Röntgen (1845–1923) allowed doctors to look straight through human flesh at bones. These X-rays were the first diagnostic images—medical pictures which reveal the cause of illness.

Today diagnostic images give doctors a very much clearer picture, and thus a better understanding of medical problems. Computerized axial tomography (CAT) scanners analyse multiple X-rays, revealing the body in depth. Other methods do not use X-rays at all. Nuclear magnetic resonance scanners use powerful magnets. Ultrasound reveals the body's internal structure with echoing noise.

▲▶ **An endoscope** lets doctors look inside the body through a tiny opening that requires little stitching afterwards. Flexible glass fibers carry a picture from the lens inside to the eyepiece. The American surgeon Basil Hirschowitz invented it in 1958.

GENETICS

Each cell in every living thing contains a threadlike chemical called deoxyribonucleic acid (DNA). Grouped in a unique order, compounds along the thread form genes—"blueprints" for growth and life. Parents pass on this genetic code to their children.

The discovery of DNA's spiral structure in 1953 was enormously important. Scientists studying it hope to devise new treatments for inherited diseases. A gene-sequencing (decoding) project that began in the 1980s aims to map human DNA.

▼ **The dark bars** on this computer print-out are gene codes.

▶ **DNA threads are spiral.** This shape is known as a "double helix."

Computing and communications

▲ Pocket calculators, invented in 1972 by Texas Instruments, replaced machines as big as typewriters.

Computing began in 1822 with spinning cogwheels. Using hundreds of gears, English mathematician Charles Babbage (1792–1871) planned a giant "Analytical Engine."

Calculating machines that ran programs and remembered results became practical only in the electronic age. Not everyone saw a need for them and even the head of IBM commented in 1947, "...there is a world market for about five computers."

Today, computer systems control everything. Many of these systems work by communicating between many computers. The first system of linked computers controled the U.S. defense missile system in the 1950s.

communications satellite

▼ Mobile telephones exchange radio signals with an antenna nearby, and regular cables carry the call from there. Bell laboratories tested the first system in 1978.

▼ The silicon wafer has hundreds of tiny integrated circuits (I.C.s) etched into it. Bob Noyce (1927–90) made one of the first I.C.s in 1959.

Types of communications
Television programs, phone calls, and the Internet allow us to communicate in different ways. T.V. and radio stations broadcast information, and millions of viewers and listeners tune in to the same programs. With phone calls, just two people exchange information privately. The Internet shares the advantages of both. Anyone can supply information, but people can choose what interests them. Advances in technology, such as mobile phones, are also changing how we communicate.

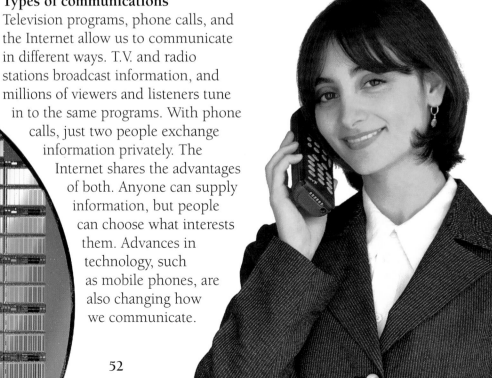

52

Computer progress

Glowing red hot, 20,000 vacuum tubes lit up the first general-purpose electronic computer. Called E.N.I.A.C., (Electronic Numerical Integrator and Computer), it was built in the U.S. in 1943–46.

signal received here

signal sent from here

It weighed about the same as a huge truck, and used so much power that the city lights dimmed when it was switched on. One of E.N.I.A.C.'s jobs was to help design the atom bomb.

The computers that followed in the 1950s also used bulky, hot, slow vacuum tubes, but by 1960 transistors were replacing them. However, computers remained huge and very expensive compared with present-day prices. The invention of the integrated circuit (I.C. or chip) changed this. A thousand transistors fitted on to an I.C. the size of an ant. With I.C.s, Digital Equipment made the first minicomputer, the P.D.P.-8, in 1963. It was as big as a refrigerator and very expensive. In the 1970s I.C.s had shrunk enough to make possible the microprocessor— a computer on a chip.

▲ **Digital telephone exchanges** use computers to route calls. Some replaced mechanical systems that had hardly changed since 1891.

▲ **Communications satellites** receive signals from a huge dish on the ground. The satellite makes the signal stronger, then transmits it to a receiving station far below. The satellites relay voice and data (information) between continents without the need for undersea cables. The U.S. Government launched the first communications satellite, SCORE, in 1958.

◄ **In-car navigation** guides motorists with a map or spoken directions. The system uses signals from satellites to judge the driver's location. The car's wheel movements and compass provide extra information. A computer combines this with maps to point the way.

Agriculture

Modern farms are increasingly using technology to produce cheap, plentiful food. Robots are now used to milk cows and computers monitor how much milk is produced. Drugs are used to protect meat animals against disease and, with the help of fertilizers and chemical sprays, farmers grow crops on what was once waste land. New machines can even control the cultivation of individual plants, monitored by an orbiting satellite.

Crops and animals themselves have changed. Not only have farmers been steadily "improving" them by selective breeding, but genetic engineers have now learned to change crop or livestock DNA directly (*see page 55*).

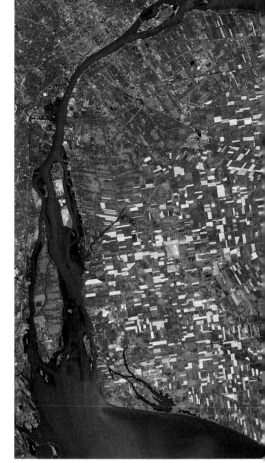

▼ A modern robot milking parlor
cleans the udders, attaches suction cups to each teat, and monitors how much milk each cow produces. Milking is one of the last of the farmer's jobs to be automated.

Livestock farming
Livestock (animal) farms are efficient food factories. Scientific breeding has doubled the number of eggs a hen lays and made cows give more milk.

Animals are kept warm, dry, and safe so they grow quickly. Electronic feeders ensure they are never hungry and drugs prevent disease.

Although this "intensive" farming cuts food costs, it can cause many animals to suffer. Rapid growth cripples some of them, and others, such as battery hens, live in unnatural, overcrowded cages.

Growing food
To harvest bumper crops, farmers use advanced machinery, fertilizers, and chemicals. None of these aids is itself new. The combine harvester, for instance, was invented in 1838, and artificial fertilizer in 1842.

However, nowadays farmers are using technology to make better use of machines and chemicals. Today's harvesters can plot where crops grow best by using satellite navigation systems. Planting only the best land saves money. Farmers can also cut costs (and protect the environment) with machines that will apply fertilizer and pesticides only where they are needed.

◀ **This electronic picture of cultivated fields** in Canada was created by an orbiting satellite. Vegetation in the fields shows up as red, and the different tones of red show crops at various stages of growth.

▶ **The Terra-Gator is a** satellite-guided crop sprayer. A global positioning satellite locates the machine in the field and tells it which fertilizers are needed to treat the different kinds of soil that are often found within the same field.

Genetic engineering

Farmers learned long ago to breed livestock from the strongest males and females in their herds. Some of the offspring inherited the genes, and thus the best qualities, of both parents. This "selective breeding" has also helped improve plants. In the 1970s scientists learned how to change genes directly. They spliced DNA from one living thing into another. This genetic engineering can even move DNA from animals to plants—or vice versa.

Genetically engineered plants may soon be able to thrive on land that is currently unsuitable for farming. Indeed, scientists have already produced a soy bean that is not harmed by weedkiller. This has encouraged farmers to use more chemicals and many people fear the weedkiller-resistant gene could spread to the weeds. They are also afraid of being poisoned themselves.

Cutting down chemicals

Fertilizers, pesticides, and weedkillers are expensive and some are dangerous poisons. Spraying can kill not only crop-damaging insects, but the predators that feed on them. Scientists are currently looking for alternatives, and natural pest control is a promising possibility. Geneticists are also developing varieties of crop plants that crowd out weeds.

▼ **Rice plants** can be scientifically bred to yield heavy crops. However, the new super-plants need expensive fertilizers and pesticides, which many farmers in developing countries cannot afford to buy.

FROST, FRUIT, AND FISH

Frost harms or kills many crops. Scientists may soon be able to reduce damage to fruits such as strawberries by using genes from fish. The flounder fish swims in water colder than 29°F. Its body does not turn to ice because its liver produces natural antifreeze. Scientists have identified the gene that gives the fish this ability and they may soon transfer it to fruit.

Building technology

As technology advances, buildings and bridges are not only becoming bigger and safer—some are actually becoming smarter, too! Towering skyscrapers and soaring bridges must stay up—even in fierce earthquakes or hurricanes. To make them safer, designers rest buildings and bridges on rubber pads to absorb the earthquake shocks, and new designs are created to add strength without increasing weight, so that they remain standing in strong winds.

Buildings are also getting taller as cities become more crowded, and with longer bridges and tunnels, railways and roads can span ever wider rivers and bays, cutting travel time. Engineers may soon make structures smarter by building fiber-optic strain gauges into them. These "glass nerves" will be able to detect corrosion or movement.

chains
concrete lump
bath of liquid silicone
"paddle" feet

▲ **The Citicorp Center in New York** is 59 floors high. To prevent it from swaying in high winds, engineers installed a tuned mass damper under the roof. This is a huge lump of concrete that weighs as much as 50 elephants, and sways in the opposite direction to the building, keeping it straight (see *inset*).

▶ **The Humber Bridge** has a central span of 1,542 yards. It was the longest in the world for more than 16 years after its completion in 1981 (the Japanese Akashi Kaikyo bridge is now the longest). To reduce weight and pressure, engineers designed the road deck of the Humber Bridge as a series of shallow boxes.

Wider buildings

In buildings made in the traditional way, pillars support the roof. But in aircraft hangars, the columns must be far apart so aircraft can pass between them. Similarly, in a sports stadium, too many pillars block the view. For such buildings engineers design light-weight roofs that can be supported by widely spaced, weaker columns. The widest spans are made like tents, and held up with steel cables anchored to the ground. Constructed in this way, Munich's Olympic Stadium covers an area equivalent to 16 football pitches.

Taller buildings

Walk up a windy hill and you'll understand the problem engineers face when they make tall buildings. The higher you climb, the stronger the wind blows. It is not difficult to make a building the wind cannot blow down, but it is hard to stop it swaying.

Big movements crack windows and make people inside feel sick. Engineers stiffen buildings by constructing them like tubes, or they fix corner-to-corner beams. A few buildings, such as New York's Citicorp Center (*see opposite*), use moving heavy weights under the roof to counteract the wind.

New materials

Plastic and carbon fiber materials could soon make longer bridges and stronger buildings possible. Steel cables cannot support suspension bridges longer than 1¼ miles, because their own weight snaps them. With lighter plastic cables, bridges could be twice the length. Engineers are experimenting with plastic and glass for strengthening concrete beams. But these materials cost more than steel, and nobody yet knows how long they will last.

▼ **The Munich Olympic Stadium** has the biggest roof in the world. It is built from a mesh of strong steel cables that hold up a light-weight covering of clear plastic panels. The German architect Frei Otto (*b.* 1925) devised its structure after studying the many strong stretched shapes that are found in nature, such as spiders' webs and soap bubbles.

▶ **Landmark Tower** in Tokyo is Japan's tallest building—it is 971 feet tall. Steel columns protect it against wind and earthquakes. The soaring complex contains shops, offices, a hotel, and the world's fastest elevators.

Power sources

The modern industrial world needs cheap, convenient energy like coal, oil, and gas. But burning these "fossil fuels" fills the atmosphere with poisons that harm the environment and change the climate—and they are not renewable so eventually supplies will run out.

Fortunately there are alternatives. Since about 1880, water stored in dams has spun turbines to generate hydroelectric power. And now scientists are learning how to extract energy from renewable, non-polluting sources such as the wind, waves, and sunlight.

▲ **A tidal barrier** dammed the Rance River in France from 1967. The tide flows out through turbines, generating enough power to serve 250,000 homes.

Power from the Sun

There is no shortage of free energy because the Sun's rays can be used to keep buildings warm and bright. It's also possible to turn sunlight into electricity. Solar collectors use mirrors to focus the Sun's rays. The point where the rays meet gets hot enough to turn water to steam. In solar power plants, the steam drives turbines which generate electric power. Photoelectric cells turn sunlight directly to electricity. German physicists Julius Elster (1854–1920) and Hans Friedrich Geitel (1855–1923) invented them in 1904. Solar cells are common in devices such as calculators, which need little power. But generating more energy requires more cells. A solar-powered car, for instance, needs to be covered in them.

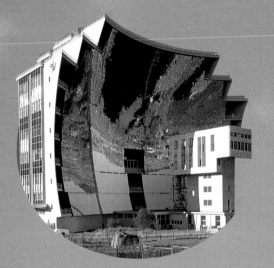

▲The Odello-Font-Tomeau solar furnace in France uses a huge array of mirrors to focus the Sun's rays. It can generate 1,000 kw of power, or heat anything at the point of focus up to 6,900°F.

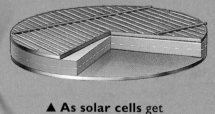

▲ **As solar cells** get cheaper and more efficient, they will contribute more to our ever-increasing power needs.

▼ **Wind farms** generate pollution-free power, but they are ugly and so noisy that they disturb people living nearby.

Wind, waves, and tide

The U.S. Department of Energy built the first large wind-power plant, at Goldendale Columbia River Gorge, Washington State, in 1982. Its three turbines produced enough electricity to supply 2,500 homes. Ten years later, wind turbines were generating 1,000 times as much power.

The sea could provide huge amounts of power, but it's hard to extract energy from waves. Tide power, though, is now widespread as it is easier to capture by damming a bay or a river mouth.

▲ **Hydrogen-powered vehicles** don't pollute the environment. However, hydrogen fuel is three times the price of petrol, and is so dangerous that robots would have to refill the fuel tank.

Power from the atom

Nuclear reactors generate electricity from radioactive fuel, such as plutonium. Heat from the decaying fuel turns water into steam, which spins turbines to power generators. Italian-American physicist Enrico Fermi (1901–1954) built the first experimental reactor in 1942. At first, it seemed to offer cheap, endless power. However, turning Fermi's experiment into a real power plant was a difficult, dangerous process. Safety fears are now so great that the building of new plants has been stopped worldwide.

Cleaner cars

The internal combustion engine (*see page 41*) is a major source of pollution, but so far there are no practical alternatives. Rather than changing the engine, scientists are looking at new, cleaner fuels, such as hydrogen. When this gas burns it produces only water, and no polluting gases or smoke. Manufacturing hydrogen takes more energy than the gas produces when it burns. However, a solar plant could produce the gas cheaply near the equator, where the Sun's rays are brightest.

▲ **Men examining a nuclear fusion reactor.** It may be possible to generate nuclear power without radioactive waste, but there are serious safety implications. Fusion is the process that heats the Sun.

GEOTHERMAL POWER

Deep below the ground, the Earth's rocks are hot. Where the heat rises to the surface it provides geothermal energy. Hot spring water heats homes in Iceland. Almost a third of the country's power needs come from this geothermal source. Where the rocks are hot enough to turn water to steam, geothermal energy can generate electricity. The first geothermal power plant began operating at Larderello, Italy, in 1904. El Salvador now produces a third of all its electricity from hot rocks.

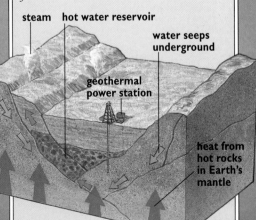

steam   hot water reservoir

water seeps underground

geothermal power station

heat from hot rocks in Earth's mantle

Everyday life

Inventions have transformed peoples' homes in many countries. People used to shop daily and cook with raw ingredients. Now convenience food comes ready-made and fridges keep it fresh. In the past, people made their own clothes and washed them by hand. Today, we buy our clothes and machines wash them. Our ancestors carried coal and water for heating, cooking, and washing, and at dusk they lit lamps. Pipes and cables now supply gas, water, and electricity. Home technology has also changed our leisure time, as inventions such as T.V. beam entertainment into our homes. And now it's changing work, too. With telephone and computer links, some people do not even have to leave home to go to work.

▲◀ **Washing machines** work while we do other things. Alva Fisher (1862–1947) devised the first in Chicago in 1906. It was not safe and many people got electric shocks.

▶ **The electric vacuum cleaner** was invented in 1901 by British engineer Hubert Booth (1871–1975). It replaced hand-pumped models like the one above. Electric vacuum cleaners help control insect pests as well as cleaning carpets.

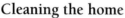

Cleaning the home

Keeping clean became easier in the late 19th century when piped water and sewers reached European homes, although many regions of the world still lack these facilities today. The invention of artificial detergents in 1917 by German chemists also helped hygiene, because detergents are better cleansers than soap and washing soda. Electrical appliances such as the vacuum cleaner and the washing machine made cleaning quicker. Artificial fabrics that did not need to be ironed—such as nylon, invented by the American textile company Du Pont in 1938—also made life easier.

The kitchen

In the past, shopping had to be done daily as not all foods could be preserved with the methods available. Nowadays, preservatives, fridges, and freezers have largely replaced the need for drying, smoking, pickling, and salting food to stop it going bad. In many modern-day households food shopping has become a weekly or a monthly event. Tinning—the 1811 invention of French chef, Nicolas Appert (1750–1841)—remains a popular method of preserving food.

Until gas cookers became cheap and popular in the 1870s, almost every kitchen had a coal-fired "range" (cooking fire) which needed regular stoking and cleaning. Electric cookers, invented in the late 19th century, began to replace gas in the 1930s, but the "Ping" of the microwave oven didn't echo from the kitchen until 1967.

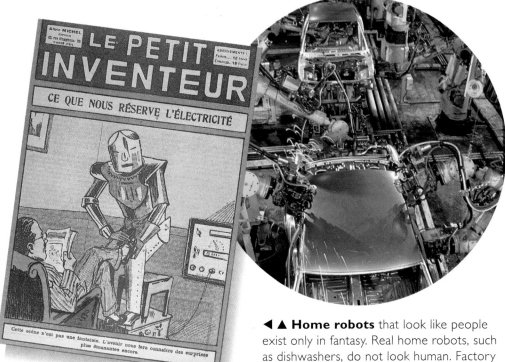

◀ ▲ **Home robots** that look like people exist only in fantasy. Real home robots, such as dishwashers, do not look human. Factory robots *(above)* are just giant arms.

TELEVISION

Scottish electrical engineer John Logie Baird (1888–1946) demonstrated a simple "Televisor" system in 1926. A spinning disk inside the set created its pictures. All-electronic televisions, without moving parts, replaced Baird's crude system in 1937, and color broadcasts began 17 years later in the U.S.

People first experimented with videotape recorders in 1956, but they were huge, heavy machines, and used about 20 miles of tape for just an hour's recording. More practical home video cassette recorders went on sale in 1975.

▼ **Television broadcasts** reach our homes by cable, from antennas on the ground, and from orbitting satellites. All three methods will soon broadcast sharper, wider, digital T.V. pictures.

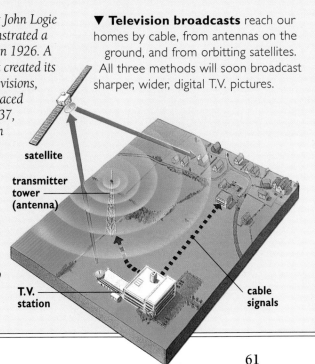

satellite

transmitter tower (antenna)

T.V. station

cable signals

Recreation and leisure
Manufactured home entertainment began with the phonograph and the radio. But neither could compete with television. From the very first broadcasts, T.V. has fascinated, educated, and entertained us, and since then a series of other inventions have made it more appealing.

Cable T.V. (1949) and satellite broadcasting (1983) have greatly increased the number of channels. Video games made television interactive in the 1970s and with the invention of camcorders in the 1980s, T.V. has now become a serious rival for the family photo album. Today, television can also connect us to a whole world of information on the Internet.

Aerospace

A sleek airliner ferries passengers across the world. It soars across the edge of space faster than the speed of sound. Not far above, the Space Shuttle carries seven astronauts into orbit.

Rocket motors made each of these achievements possible. They powered the first high-speed plane, and they lifted the first space probe, *Sputnik I*, into orbit. These record-breaking flights, both in 1947, helped create the aerospace (aircraft and spacecraft) industry.

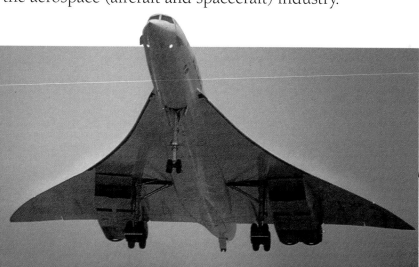

◀ ▼ **The Space Shuttle,** which first flew in 1981, is launched like a rocket. It gathers valuable information about space. Here (*left*) a telescope is seen emerging from the cargo hold.

Supersonic flight
Rockets powered the first aircraft to travel at supersonic speed. The experimental American Bell X-1 aircraft flew at 670 mph in 1947. Supersonic fighter aircraft powered by jet engines followed. By 1976 Britain and France had developed Concorde, the first supersonic passenger aircraft.

▲ **Anglo-French Concorde** transports its 128 passengers twice as fast as the speed of sound. The flight from London to New York takes under 3½ hours.

Exploring space

Launching a spacecraft requires a powerful rocket to overcome gravity.

In 1947 the Soviet Union launched the first satellite, *Sputnik I*, on an SS-6 missile, powered by kerosene and liquid oxygen. Liquid-fueled rockets of this type were invented in 1926 by American physicist Robert Goddard (1882–1945).

Less than four years after the launch of *Sputnik*, Russian Yuri Gagarin (1934–68) became the first astronaut to be launched into space, and in 1969 the *Apollo 11* mission landed an American crew on the Moon. Four successful moon missions followed.

Today the space probes that go beyond Earth's orbit do not carry crews because uncrewed missions do not risk human life. Astronauts from the United States and Russia still work in Earth's orbit, though. Once rivals in the "Space Race," the two nations are now planning an orbiting space station together. By cooperating, they are sharing the enormous cost of space science.

Space war and peace

The space programs of the U.S. and the Soviet Union began when each country felt threatened by the other. This initial research into nuclear missiles supplied the rockets that lifted astronauts into orbit.

Space technology is now used primarily for peaceful purposes, such as satellites for navigation, television, and telephone.

LIVING IN SPACE

While American astronauts were landing on the Moon, Soviet crews were busy designing and building a series of space stations to be used in the Earth's orbit. The experience they gained will be vital in the future, as there are plans to construct an international space station like the one on the left. When complete, the station will be a huge, permanent orbiting scientific base.

▼ **The Stealth B2 bomber,** built in the United States by the Northrop Company, uses special materials and a smoothly curved shape to reduce radar reflections and hide from the enemy.

▼ **The X-30 "spaceplane"** was designed to burn fuel in air or liquid oxygen, so it could reach orbit without rocket boosters. However, it proved too costly and never actually flew.

Timetable of Inventions

What was being invented in 1891? This list provides some answers. Inventions that are described or mentioned earlier in the book are marked with a black dot (•).

Because this is only a list, the date shown is the most important one. Giving the date, though, is not always simple. We know the dates of some inventions (especially ancient ones) only very approximately. For example, we list the invention of the helicopter as 1907, because this is the year that the first crude machine lifted off the ground. But Leonardo da Vinci sketched helicopters in 1486; and they became a practical form of transport only in 1936.

The list ends in 1990 because with each year that passes it becomes more difficult to identify truly original inventions. Many of the new machines that make our lives easier are simply new combinations of old ideas.

B.C.

• 2000000	Tools
• 600000	Fire
400000	Spear
• 250000	Ax
50000	Oil lamp
45000	Paint
• 40000	Boat
• 30000	Bow and arrow
20000	Needle
13000	Harpoon
10000	Hammer (stone)
10000	Fishing net
8000–6000	Brewing
• 7000	Spinning
• 7000	Pottery
6000–4500	Fish hook
5500	Metalworking
5000	Brick
• 5000	Basketwork
• 5000	Weaving
• 5000	Sled
• 4000–3000	Sailing vessels
4000	Rope
4000	Writing
3500	Balance, scales
• 3500	Bronze
3500–3001	Mirror
3500	Nail
3500	Papyrus
• 3500	Plow
3500	Potters' wheel
• 3500	Cart wheel
3200	Ink
3000	Button
• 3000	Arch
3000	Fish hook, barbed
• 3000–2500	Glass
• 3000	Loom, framed
3000	Plumb line
• 3000	Measurement, standard length
3000–2500	Soap
3000	Wig
2950–2750	Dam
2800	Calendar, lunar
2700–2000	Sewers
2500	Baths
• 2500	Iron smelting
• 2500	Cement
2500	Perfumes and deodorants

2500	Skis, snow
• 2500	Standard weights
2350	Lavatory
2300	Embalming
2000	Ball
• 2000	Chariot
2000–1800	Harness
2000	Locks and keys
2000	Shoe
• 2000	Spoked wheel
1800	Medical instruments
• 1500	Shaduf
• 1500	Bottle, glass
1400–1200	Sling, slingshot
1350	Welding
1325	Printing from engraved surfaces
1250	Knitting with needles
1100	Kite
1100	Parasol
1000–700	Sundial
700	False teeth
690	Aqueduct
620	Coins
600	Cast iron
600	Magnet
• 600–300	Merkhet and groma (surveying instruments)
• 600	Winch, windlass, capstan
550–510	Map
500–400	Carpet
450	Abacus
• 450–400	Pulley
400	Catapult
• 300	Gears (cogwheels)
• 300–250	Water clock
285	Lighthouse
• 236	Archimedean screw
• 200	Concrete
• 190	Vellum
150	Screw press
100	Crossbow
100	Glass blowing
• 100–0	Water wheel

A.D.

50	Horseshoe
• 50	Vertical water wheel
25–220	Saddle
• 50	Magnifying lens
• 50	Steam turbine
• 105	Paper
124–128	Dome
• 200	Wheelbarrow
200	Ice skates
600	Chess
600–700	Money, paper
650	Scissors
• 740	Printing of text from wood blocks
750	Iron, laundry
• 750	Rudder
840–1000	Musical notation
• 850	Gunpowder
868	Printing, book
• 900	Windmill
• 900	Horse collar and harness
969	Playing cards
983	Canal lock
1000	Lens, glass
• 1000	Spinning wheel
• 1040–1050	Printing, movable type
1071	Fork
• 1086	Camera obscura
• 1090	Compass, magnetic
• 1094	Mechanical clock
1103	Fireworks
• 1250	Astrolabe
• 1250	Jack, screw
• 1280	Cannon
1286	Spectacles
1291	Mirror, glass
1298	Longbow
• 1350	Hand cannon
1350–1400	Alarm clock
• 1400	Oil paint
1400	Returning boomerang
1421	Explosive shell
1430	Sunglasses
1436	Perspective
1440	Engraving

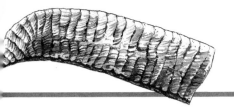

• 1450	Harquebus (gun)	• 1769	Steam engine,	• 1829	Internal	• 1860	Milking machine
1450	Concave lens		separate		combustion	1861	Plastics (Parkesine)
• 1450	Printing press		condenser		engine, gas	1861	Postcard
1498	Toothbrush	1769	Steam carriage		powered	1862	Pasteurization
1504–1511	Watch	1770	Pencil eraser	1829	Sewing machine	1863	Trains,
1509	Wallpaper	1777–1786	Threshing and	1830	Lawn mower		subway
1525	Rifle		winnowing	1831	Electric bell	1865	Antiseptics
1528	Grenade		machines	1831	Dynamo	1865	Riveting machine
1530	Bottle cork	• 1779	Bridge, iron	1831	Electric motor	1865–7	Dynamite
1538	Diving bell	• 1782	Steam engine,	1832	Tram	1866	Torpedo
1550	Nuts, bolts,		rotary	1834	Railway signals	1867	Reinforced
	spanners	• 1783	Balloon, gas	• 1835	Photographic		concrete
1565	Pencil	• 1783	Balloon, hot air		negative	1868	Stapler
• 1569	Map, Mercator	1783	Parachute descent	• 1838	Combine harvester	1871	Margarine
	projection	• 1784	Factory	1838	Morse code	1871	Wind tunnel
• 1575	Cross-staff		automation	• 1839	Bicycle	1872	Chewing gum
1589	Knitting machine	1784	Bifocal spectacles	1839	Steam hammer	1872	Monorail
• 1590	Compound	1785	Gas lighting	• 1840	Interchangeable	1867–8	Barbed wire
	microscope	1785	Lifeboat		parts	1874	Electric car
• 1593	Thermometer	1785	Power loom	1840	Postage stamp	1874	Jeans
1596	Water closet	1791	Coal Gas	• 1842	Fertilizer, artificial	1874	Typewriter
• 1608	Telescope	1798	Litho printing	1843	Christmas card	• 1876	Telephone
1609	Thermostat	1798	Mass production	1844	Anaesthetics	1877	Fingerprinting
1622	Slide rule	1800	Battery	• 1844	Electric telegraph	1877	Cylinder record
• 1624	Submarine	• 1802	Gas cooker	• 1845	Pneumatic tire		player,
1637	Rain umbrella	• 1802	Railway	1845	Tarmacadam		(phonograph)
1644	Barometer		locomotive	• 1847	Arc lamp	1878	Electric street
1654	Vacuum pump	• 1803	Steam-boat,	1848	Yale lock		lighting, arc
• 1656	Pendulum clock		paddle-wheel	1849	Dry cleaning	• 1878	Microphone
1662	Omnibus	1806	Gas, public supply	1849	Refrigeration	1879	Cash register
• 1670	Reflecting	• 1807	Gas lighting,	1849	Safety pin	1879	Electric train
	telescope		street	1850	Match, safety	1879	Saccharin
• 1675	Precision	• 1811	Canned food	1850–1860	Ballbearing	1879	Toilet paper,
	microscope	1815	Miners' safety lamp	• 1852	Airship		perforated
1686	Ice cream	• 1815	Macadamized	1852	Bee hive	• 1880	Electron tube
1698	Steam pump		roads	1852	Bottle, screw top	1880	Seismograph
1701	Seed drill	1816	Stethoscope	1853–1856	Ambulance	1881	Gas fire
• 1709	Iron smelting,	1818	Revolver	1853	Condensed milk	• 1881	Hydroelectric
	coke	1818	Tunneling	• 1853	Glider		power station
1700	Piano		machine	1853	Hypodermic	1881	Streetcar (electric
• 1712	Steam engine	1820	Elastic		syringe		tram)
1718	Machine-gun	1820	Electro-magnet	1854	Elevator	• 1881	Electric light bulb
• 1731	Quadrant	• 1820	Iron ship	1855	Can opener	• 1882	Electric iron,
	(navigational	1821	Electric motor	1856	Steel		laundry
	instrument)	• 1822	Programmable	1854	Vacuum tubes	• 1883	Electric fan
1752	Lightning		computer	1859	Rechargeable	1883–1885	Machine gun
	conductor	• 1824	Pin-making		battery	1884	Fountain pen
1757	Sextant		machine	1859	Internal	• 1884	Gasoline engine
1760	Roller skates	• 1824	Portland cement		combustion	1884	Steam turbine
1762	Sandwich	• 1826	Photography		engine	• 1885	Motor car
• 1769	Spinning machine,	1827	Contact lens	1859	Oil rig	• 1885	Motorcycle
	water powered	1827	Water turbine	1860	Linoleum	1886	Coca Cola

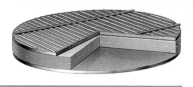

• 1888	Motion pictures	1906	Sound radio	1942	Missile	1964	Word processor
• 1888	Inflatable cycle		broadcasting	• 1942	Nuclear reactor	• 1967	Satellite
	tire	• 1907	Helicopter	1943	Guided missile		navigation
• 1888	Disk record player	1907	Fax	1945	Atomic bomb	1970	Floppy disk
	(gramophone)	1908	Geiger counter	• 1945	Electronic	1971	Microprocessor
1889	Agricultural	1910	Aircraft carrier		computer	1971	Space station
	tractor	1910	Food mixer	• 1946	Microwave oven	• 1972	CAT scan
1889	Coin-operated	1912	Supermarket	• 1947	Supersonic flight	• 1972	Calculator, pocket
	telephone	1913	Assembly line	1947	Instant	• 1972	Home video game
• 1891	Automatic	1913	Crossword puzzle		photography	1973	Skateboard
	telephone	1913	Stainless steel	• 1947	Transistor	• 1975	Video cassette
	exchange	1914	Zip fastener	• 1949	Cable T.V.		recorder
• 1891	Electric oven	1914	Traffic light	1950	Credit card	1976	Disposable diaper
• 1891	Hang glider	1915	Air conditioning	• 1950	Diagnostic	• 1976	Supersonic
1891	Electric kettle	1915	Heat resistant		ultrasound		passenger flight
1892	Diesel engine		glass (Pyrex)	1952	Artificial heart	• 1978	Cellular
1892	Electric fire	• 1917	Detergent		valve		telephone
1892	Vacuum flask	1918	Sonar	1953	Heart-lung	• 1977	Personal
1893	Dishwasher	1920	Hair dryer		machine		computer
1894	Concrete bridge	1921	Lie detector	• 1954	Freeze-dried food	1979	Personal stereo
1894	Escalator	1921	Motorway	1954	Organ transplant		(Walkman™)
1895	Movie projector	• 1923	Electric home	• 1954	Atomic power	• 1979	Spreadsheet
• 1895	Inflatable car tire		refrigerator		station	1980	Computer
• 1895	Radio	1923	Planetarium	1954	Transistor radio		networks
• 1895	X-rays	• 1924	Frozen food	1954	Nuclear	1980	Halogen lamp
1896	Cotton-picking	1925	Water-skiing		submarine	1980	Supermarket
	machine	• 1926	Liquid fueled	1954	Color television		checkout
1896	Electric hand-drill		rocket	1955	Non-stick pan		scanner
1897	Aspirin	• 1926	Television	1956–1958	Heart pacemaker	1981	Airbag
1898	Magnetic sound	1926	Pop-up toaster	1956	Transatlantic	• 1981	Space Shuttle
	recording	• 1928	Antibiotics		telephone cable	• 1981	Stealth fighter
1898	Telephone	1928	Electric razor	• 1956	Videotape	1982	Artificial heart
	answering	1928	Sticky tape	• 1957	Food irradiation	• 1982	Wind farm
	machine	1929	Yo-Yo	• 1957	Robot	1983	Compact disk
1900	Submarine	1930	Sliced bread	• 1957	Space satellite	1983	Orbiting infrared
	periscope	1931	Electron	• 1958	Endoscope		telescope
1900	Sound movie		microscope	1959	Car seat belt	1983	Portable
1901	Hearing aid	1933	Radar	1959	Hovercraft		computer
1901	Instant coffee	1935	Parking meter	• 1959	Integrated circuit	• 1983	Satellite T.V.
1901	Razor blades	• 1935	Color film	1960	Laser	1984	Genetic
• 1901	Electric vacuum	1935	Tape recorder	1960	Plastic hip		fingerprinting
	cleaner	1937	Ball-point pen	1961	Mouse, computer	1985	Post-its™
1902	Tea bags	1937	Jet engine		pointing device	1985	Scanning
• 1903	Propeller	1937	Radio telescope	• 1961	Space travel		tunneling
• 1903	Aircraft	• 1938	Nylon	• 1962	Communications		microscope
1903	Soap powder	1938	Photocopying		satellite	1986	Keyhole surgery
• 1904	Photoelectric cell	1939	Fluorescent	1962	Skateboard	1987	Digital audio tape
• 1904	Geothermal		lighting	1963	Carbon fibre	1987	Gene gun
	power station	1939	Jet aircraft	1963	Cassette tape	1987	Laser scalpel
1906	Jukebox	1940	Pin-ball machine		recorder	1989	Game Boy™
• 1906	Electric washing	1940	Suntan lotion	• 1963	Minicomputer	1990	Laser dental drill
	machine	1941	Aerosol can	1963	Ring-pull cans	1990	Space telescope

Glossary

Most words dealing with inventions are explained in the text. So try looking in the index if you cannot find the word you are looking for here.

alloy A mixture of metals.

antibiotic A drug that kills or slows the growth of harmful germs.

architecture The art and science of building construction.

artisan A craftworker who makes things by hand.

automation The replacement of a machine that needs human control with a machine that controls itself.

capstan A useful, drum-shaped winding machine that turns on an upright support.

cartographer A map maker.

casting Pouring a liquid into a mold so that it assumes the shape of the mold when it solidifies.

cement Powdered building "glue" that sets when mixed with water.

cogs The raised teeth around the edge of a gear wheel.

communications The transmission of information, such as writing, speech, or numbers.

compass A magnetized north-pointing needle used for navigation.

compound microscope A microscope with two or more lenses.

concrete A mixture of sand and stones, held together by cement.

cylinder The smooth-sided tube of a pump or motor in which a piston slides (*see piston*).

DNA Deoxyribonucleic acid—a thread-like chemical that contains genes (*see gene*).

draught animal An animal such as a horse or ox used to pull a load.

filament A thin strand of material inside a light bulb that glows with light when heated by electricity.

gear wheel A wheel fitted with pegs or cogs that allow it to turn another similar wheel.

gene A portion of DNA through which living things pass on one particular feature to their offspring — such as hair color in humans.

genetic engineering The artificial changing of genes.

gravity The invisible force that pulls all objects toward each other.

hull The body or frame of a ship or boat.

hydroelectric power Electricity generated from falling water.

Industrial Revolution A time of rapid development in the late 18th and 19th centuries associated with machine production.

laser A special source of light that gives out an intense beam of just one color.

lever A rotating arm or bar used to make lifting easier, or to make a small movement larger.

lintel A level bar supporting the top of a door or window frame.

locomotive An engine on wheels that can move itself and pull a load.

loom A frame used for weaving.

mass production Making large numbers of similar items from standard parts in a factory.

Middle Ages The period of European history between ancient and modern times: approximately A.D. 500–1500.

mold A container used to hold liquid or soft materials that take on its shape when they solidify.

orbit The circular path that a satellite follows.

organic farming Farming using traditional methods, or without chemicals.

piston A sliding block that moves inside the cylinder of a pump or motor.

plumb line A weighted line used to check that walls are straight.

preservative A substance that stops things rotting.

pulley A grooved wheel around which a rope runs, changing the direction of a pulling force, or making it easier.

Renaissance A period of European history, from the 14th to 16th centuries in which there was renewed interest in classical arts.

research The investigation and study of a particular subject, often to solve a problem.

satellite An object in space that circles another object in the same way that the Moon circles the Earth.

simple microscope A microscope with just one lens.

smelting Heating rocks to extract and purify the metal they contain.

technology Knowledge or science put to practical use.

telegraph A signaling system that allows two or more stations to communicate using a code.

turbine A wheel, fitted with slanting blades, that spins in a jet of water, air, or steam.

vacuum A space from which even the air has been removed.

vacuum tube A glass tube that makes electricity flow in one direction only.

winch A useful, drum-shaped winding machine that turns on a level support.

yarn Thread used in weaving.

Index

PICTURE ACKNOWLEDGMENTS

ARTWORK

Peter Bull, 47; Harry Clow, 8 (top), 9 (top); Mark Franklin, 13 (top), 21 (right), black icons; Richard Hook (Linden Artists), 8 (left), 13 (bottom right), 15 (top center), 19 (bottom), 22 (bottom left), 23 (top right), 25 (bottom); John James (Temple Rogers), 12 (bottom), 22 (bottom right), 25 (top); Janos Marffy, 8–9 (center); Keith Page (Temple Rogers), 34 (center), 35 (center), 35 (center right); Simon Raulstone, 9 (bottom); Michael Roffe, 40–41 (bottom), 42–43 (center); Christopher Rothero (Temple Rogers), 18 (bottom); Peter Sarson/Richard Chasemore, 10–11 (center and right), 14–15, 17, 18 (top), 18 (bottom left), 19 (center right), 20 (left), 23 (top left), 28–29, 30–31 (center), 36 (top left), 37 (top), 38–39, 40 (top left), 40 (bottom left), 41 (top right), 44–45, 46, 55, 57, 61; Peter David Scott (Wildlife Art Agency), 10 (left); Guy Smith (Mainline Design), 49, 50–51, 52–53 (center), 56, 63 (bottom); Roger Stewart, 20–21 (center), 32–33, 36–37 (center), 41 (center right), 59 (bottom right); Michael Welply, 10–11 (top), 12–13 (center), 26–27, 30 (left); Michael White (Temple Rogers), 34–35 (bottom).

PHOTOGRAPHS

t = top, b = bottom, c = center, l = left, r = right

Endpapers: Zefa Pictures; 4 Sandia National Laboratories/Science Photo Library; 6 The Cadbury Collection, Birmingham Central Library; 7 Chas Wilder; 8 John Barlow; 9 John Barlow; 13 Christine Osborne Pictures; 14/15 A.C. Tidswell/Tony Stone Images; 15 t Dallas Museum of Art, USA/Werner Forman Archive, b Jorn Stjerneklar/Impact Photos, t Peter Newark's Western Americana; 16 c Ralph Pitchford, b Roberta Parkin/Impact Photos; 17 Sonia Halliday Photographs; 18/19 B & C Alexander; 19 Comstock; 20 Sonia Halliday Photographs; 22 l Bulloz, r Joseph Needham, 'Science and Civilisation in China', vol. 4 part II, Cambridge University Press 1965; 23 Sonia Halliday Photographs; 24 t Musee des Beaux-Arts, Valenciennes/Giraudon/The Bridgeman Art Library, b Robert Harding Picture Library; 26 l Ontario Science Centre, r Image Select; 27 l Bibliotheque de l'Institut de France/AKG London, r National Gallery, London/The Bridgeman Art Library; 28 b Royal Astronomical Society; 31 t The Cadbury Collection, Birmingham Central Library, bl Sonia Halliday Photographs, br British Library, London/The Bridgeman Art Library; 33 t The Ironbridge Gorge Museum Trust, b Robert Harding Picture Library; 37 Manchester City Council; 38 l Corbis-Bettmann, c Christie's, London/The Bridgeman Art Library,; 39 Mary Evans Picture Library; 41 Getty Images; 42 Getty Images; 43 TRH Pictures; 44 Science and Society Picture Library; 45 t Mary Evans Picture Library, c E.T. Archive, b Mary Evans Picture Library; 46/47 Mary Evans Picture Library; 47 The National Archives/Corbis; 48 l Science and Society Picture Library, r The Royal Photographic Society, Bath; 49 l Chas Wilder, r UPI/Corbis-Bettmann; 50 Corbis-Bettmann; 50/51 John Greim/Science Photo Library; 51 t TRH Pictures, c Sinclair Stammers/Science Photo Library, b Dr K.F.R. Schiller/Science Photo Library; 52 t Chas Wilder, bl Manfred Kage/Science Photo Library, br John Barlow; 53 l Sittler Jerrican/Science Photo Library, r Zefa Pictures; 54 Farmers Weekly; 54/55 CNES, 1988 Distribution Spot Image/Science Photo Library; 55 Farmers Weekly; 56 t Nikko Hotels/The Stubbins Associates Design Architects, in association with Mitsubishi Estate Company Ltd, b Leslie Garland; 57 Edward Jacoby/The Stubbins Associates; 57 (inset) H. Zeyss/Zefa Pictures; 58 l Zefa Pictures, r Zefa Pictures; 58/59 Nick Wood/Robert Harding Picture Library; 59 l Mercedes-Benz (UK) Ltd, r Roger Ressmeyer, Starlight/Science Photo Library; 60 tl Mary Evans Picture Library, bl Hoover European Appliance Group, tr Hoover Ltd, br Mary Evans Picture Library; 61 l Mary Evans Picture Library, r Zefa Pictures; 62 t TRH Pictures, b Zefa Pictures; 62/63 NASA/Science Photo Library; 63 t David Hardy/Science Photo Library, b TRH Pictures.

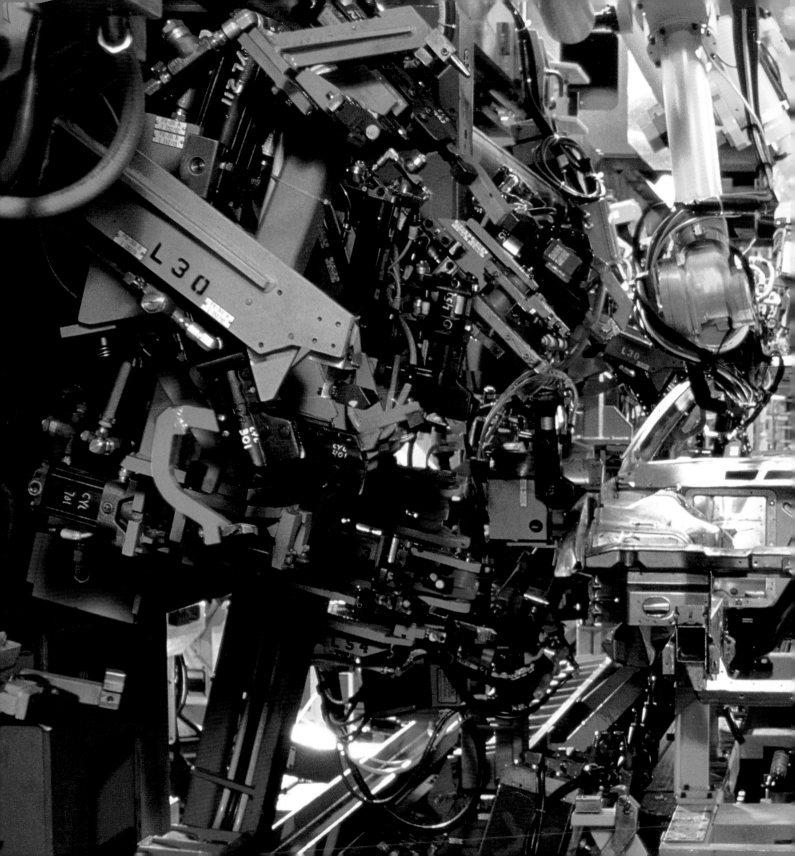